Wom

Women's rights and bioethics

Edited by
Lorraine Dennerstein

With contributions by
Margaret Baltes
Rosie Beaumont
Rebecca Cook
Susan Feldman
Amira Frljak
R. M. den Hartog-van Ter Tholen
Monika Hauser
H. Patricia Hynes
Attiya Inayatullah
Michèle Jean
Penny Kane
Genoveva Keyeux
K. Kusum
Sister Christa Mary-Jones
Elizabeth McGregor
Roland Edgar Mhlanga
Lucile Newman
Mihr Pjskic
Rika Pretorius
Silvina Ramos
Janice G. Raymond
Berit Schei
Laila Shukry El-Hamamsy
Elisabeth Steinhagen-Thiessen
Nahid F. Toubia

ETHICS

UNESCO PUBLISHING

The 'Ethics' series is directed by Georges Kutukdjian,
Director a. i. Division of Human Sciences,
Philosophy and the Ethics of Science and Technology.

The designations employed and the presentation of the material in this document do not imply the expression of any opinion whatsoever on the part of the UNESCO Secretariat concerning the legal status of any country, territory, city or area or of their authorities, or concerning the delimitation of their frontiers or boundaries.
The ideas and opinions expressed are those of the contributors to this report and do not necessarily represent the views of the Organization.

Published in 2001 by the United Nations Educational, Scientific and Cultural Organization
7, place de Fontenoy, 75352 Paris 07 SP
Typeset by : Gérard Prosper
Printed by : Corlet, Imprimeur, S.A.
14110 Condé-sur-Noireau (France)
N° d'Imprimeur : 1389
Dépôt légal : mars 2001

ISBN 92-3-103765-X

Printed in France

Preface

Bioethics cannot ignore the impact of the use of biomedical technological progress on the status of women. Without any doubt, this progress opens up vast prospects for improving women's well-being and health. However, it can also give rise to new forms of discrimination or constraints, for example in the field of medically assisted human reproduction or in prenatal diagnosis, in particular with the possibility of selecting an embryo according to gender.

Four international conferences have dealt with issues relating to the status of women: the International Conference on Population and Development (Cairo, 1994), the World Summit for Social Development (Copenhagen, 1995), the Fourth World Conference on Women (Beijing, 1995) and the World Conference on Science (Budapest, 1999).

The Fourth World Conference on Women adopted a Declaration which, in Paragraphs 30 and onwards, underlines the fundamental rights of women and highlights the link between progress in life sciences and issues which directly concern the health of women, particularly in terms of access to health care and human reproduction.

Organized under the aegis of UNESCO and the International Council for Science (ICSU), the World Conference on Science focused on several questions related to women's health and status and underlined in particular the need for women to be involved in all aspects of science and technology.

Both the Declaration on Science and the Use of Scientific Knowledge and the Science Agenda – Framework for Action, adopted by the Conference, make reference to the role of women in this field and call for special efforts to be made in order to ensure the full participation of women and girls in all aspects of science and technology.

In Point 3, 'Science for Development', the Declaration states that it is essential that the fundamental role played by women in the application of scientific development to food production and health care be fully recognized, and calls for major efforts to be made for women to have a better understanding of scientific advances in these areas.

The International Bioethics Committee (IBC) of UNESCO has always taken account of the impact of progress in the life sciences on the status of women. At its Fourth Session, in October 1996, the IBC organized a round table on the topic 'Bioethics and Women'. Following this round table, and recognizing the need for a study of women's health within the perspective of human rights and within the framework of bioethics, the IBC established a Working Group to prepare a report on this topic. The draft report was presented at the Fifth Session of the IBC (Noordwijk, Netherlands, 1998) by the Chairperson of the Working Group, Attiya Inayatullah (Pakistan) and the Rapporteur, Lorraine Dennerstein (Australia), and was the basis for the present book.

This book reminds us, if need be, of the scale of the problems. It encourages us, men and women, to fight for the dignity of women or else forfeit the right to pronounce the words 'human dignity'. With its comprehensive scope and concrete approach, it includes the concept of diversity, one of the fundamental aspects of bioethics. The eminent specialists who contributed to this book and who are committed to promoting the rights of women should be thanked for their dedication.

Ryuichi Ida
Chairperson,
International Bioethics Committee, UNESCO

Contents

Women's health — the context: Recent recommendations of United Nations conferences and initiatives of the Commonwealth of Australia

Lorraine Dennerstein*

Women's health

There is a great deal of evidence to show that the health experiences of women are different from those of men, at all ages and in all societies (Kane, 1991). The women's health movement, particularly the community-based women's health groups of the 1970s and 1980s, drew attention to difficulties women experienced with health systems (Commonwealth Department of Community Services and Health, 1989). Individually and collectively, women asked to be treated with dignity and respect, to be given information to enable them to make decisions and to become equal partners in their health care. They also asked to have input into health-planning and research. Although women form the majority of health workers, they were often inadequately represented in the decision-making process and were thus unable to influence the allocation of resources or the style of health-care systems (Dennerstein et al., 1989). Analyses of research funding in many

* Director, Office for Gender and Health, Department of Psychiatry, University of Melbourne (Australia).

developed countries found that only a small minority of such funds addressed issues that women thought were important to their health (United States National Institutes of Health, 1992). Other systematic ways in which women were disadvantaged by health systems included their exclusion from epidemiological studies and clinical trials (Freedmen and Maine, 1993).

Figure 1. Life expectancy at birth for Aborigines and inhabitants of selected countries (1985)

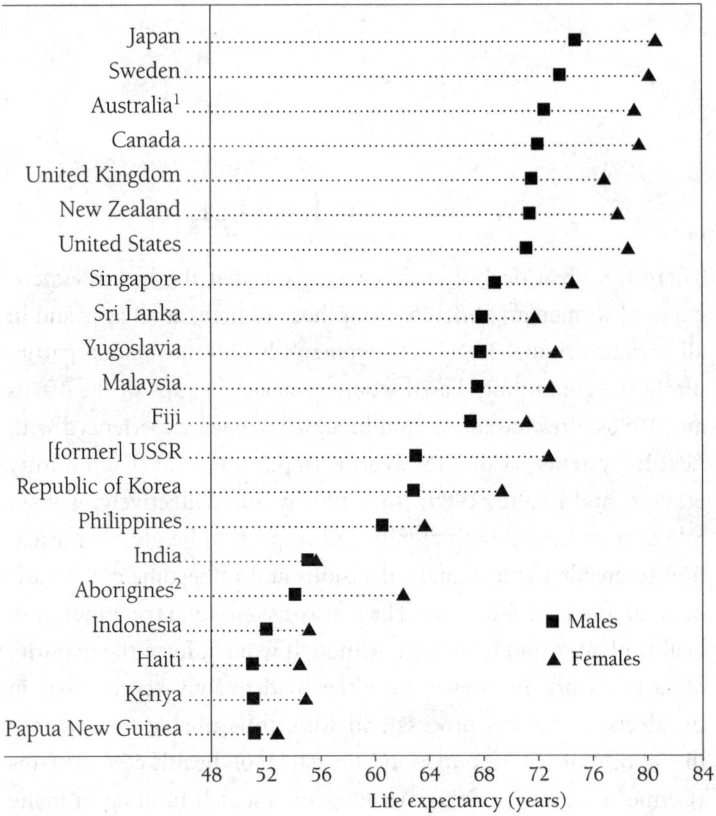

1. Includes [Aus.] Aborigines and Torres Strait Islanders.
2. Includes Torres Straits Islanders.

Source: United Nations, 1987; Australian Bureau of Statistics 3302.0. Quoted in *Australia's Health*, 1990.

But gender is not the only adverse factor. Certain social factors such as race, class and poverty have substantial effects and there are complex relationships between these factors and gender. For example, life expectancy for Aboriginal and Torres Strait Islander people is considerably lower than for the rest of the Australian population, resembling that of women in less economically advantaged countries (see Figure 1).

In 1989–1991 the life expectancy for Aboriginal and Torres Strait Island females in Western Australia was 62 years, compared to 81 years for all females. This reflects a higher mortality rate among Aboriginal and Torres Strait Islander women. After age 35 the female mortality rate of Aboriginal and Torres Strait Islander women was more than three times higher than that of other women worldwide (see Figure 2). There is considerable diversity among women, and discrepancies in women's health outcomes – between women in different countries and between different groups of women in the same country – reflect factors such as age, race, class, educational background and functional capacity.

Figure 2. Aboriginal and Torres Strait Islander death rates[1] *(1990–1992)*

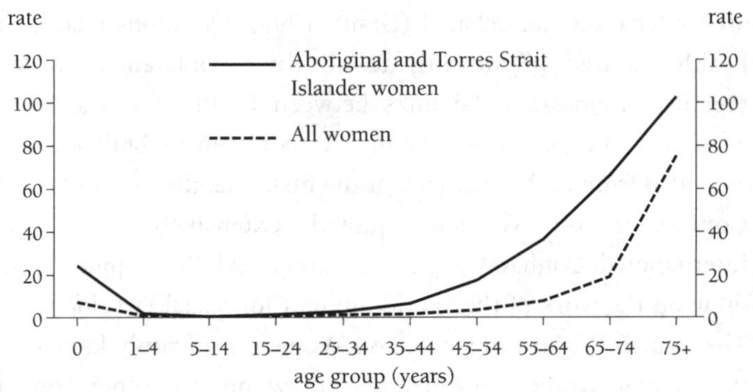

1. Deaths per 100,000 females.

Source: Australian Institute of Health and Welfare, *Australia's Health*, 1994.

These findings have led to an increasing awareness of the need for a different approach to health care, an approach that recognizes the particular circumstances of women's lives. Such an approach would involve a broader understanding of how general health is affected by social, cultural, economic, political, nutritional and lifestyle factors. There is growing recognition of the need for an interdisciplinary and inter-sectoral approach that would bring together knowledge and methodology from a wide range of discipline: health and social sciences, education, engineering, law and economics.

International activities

Against the backdrop of these concerns, we have witnessed a growing international collaboration in women's health politics in recent years, with an agenda to improve women's health everywhere. Three United Nations conferences held in the space of a twelve-month period – the International Conference on Population and Development (Cairo, September 1994), the World Summit for Social Development (Copenhagen, March 1995) and the Fourth World Conference on Women (Beijing, September 1995) – changed the way issues such as development, population, poverty, women's reproductive health and human rights are understood and debated (Grant, 1995). Questions related to health featured prominently at all three conferences, which together emphasized the links between health, education, the economy and the environment. The conferences built on and extended from each other so that the final one, the Fourth World Conference on Women, quoted extensively from the International Conference on Population and Development and drew on the work of the World Summit for Social Development. The Fourth World Conference on Women, commonly known as the Beijing Women's Conference, drew on two other United Nations conferences: Environment and Development, held in Rio in 1992, which recognized the key role of women in preserving and managing the environment; and Human Rights, held in Vienna in 1993, which integrated violence against women and other women's human-rights issues into the overall United Nations human-rights agenda and activities.

Each conference underlined the link between the disadvantages women experience in health issues and the continuing inequities in women's general status in society. It was observed that women remain disadvantaged in all socio-demographic measures of equity, and that socio-economic disadvantage, combined with a lack of autonomy or control over their bodies and lives, are major factors in high levels of female morbidity and mortality. There was recognition that while much of the burden of women's ill health relates to the general state of developing countries, many of the same issues are faced by women in developed countries.

International Conference on Population and Development (Cairo, 1994)

The International Conference on Population and Development (ICPD) was established by a resolution of the United Nations Economic and Social Council in 1989 to consider population issues within the context of national goals for sustained economic growth, sustainable human development and improved education and economic status for women. Thus, ICPD was given a much broader mandate than previous population conferences, i.e. those held in Bucharest in 1974 and in Mexico City in 1984, and this enabled a shift from family-planning issues to reproductive-health issues examined within the context of sustainable development.

The Programme for Action concerning approximately fifty issues included the following topics: the relationship between population, sustained economic growth and sustainable development; gender equality and the empowerment of women;

structure, rights and roles of the family; population growth and structure; reproductive rights and health; health, morbidity and mortality; internal population distribution; population, development and education; technology, research and development; and partnerships with non-governmental organizations, or NGOs (United Nations, 1994*b*).

The major health focus was on the shift to a new understanding of women's reproductive health, defined positively as a state of complete physical, mental and social well-being, and not negatively, as in the past, as merely the absence of disease or infirmity. The assertion was made that in those countries where abortion was legal the practice should be safe, but that the focus should be on prevention rather than abortion. The role of women as *primary custodians of family health* was given attention. There was also emphasis on women's health in relation to safe motherhood and HIV/AIDS, where the need for sex education and information was stressed, as was the need to encourage men to participate and assume responsibility.

The section on women's empowerment argued in favour of improving the status of women, calling for increased education, skills development and employment opportunities, a reduction in discrimination and a respect for equal rights in relation to landholding and inheritance for women. Government action was demanded in order to reduce the occurrence of forced marriages, excision (often called female genital mutilation), and prenatal sex selection. Also reaffirmed were the reproductive rights of couples and individuals to decide freely the number of children they would like to have and the time between each pregnancy. The corresponding report of the Conference (United Nations, 1995*b*) calls for specific policies to eliminate all forms of violence in society and intra-family discrimination against girls with regards to health.

World Summit for Social Development (Copenhagen, 1995)

The objectives of the World Summit for Social Development, as set out in Resolution 47 of the 1992 meeting of the United Nations General Assembly, were the promotion of higher standards of living, full employment and the effort to find solutions to international problems relating to society in general, and to the economy and health in particular.

This summit meeting placed people at the centre of all development, and it published a declaration containing ten commitments forming a new social contract at the global level. Among the actions covered that are vital for advancing women's health were the following: creating the legal environment that would enable people to achieve social development; achieving equality and equity between women and men, as well as recognizing and enhancing women's participation and leadership; attaining universal and equitable access to quality education; attaining access for all to primary health care as well as the highest standard of physical and mental health; developing community-based health-education programmes, particularly for women and girls; and expanding integrated and inter-sectoral approaches when providing for the protection and promotion of health for all through the process of economic and social development. Targets for a reduction in maternal mortality were also set, along with a commitment to make primary reproductive health care available to all by 2015.

Fourth World Conference on Women (Beijing, 1995)

The Beijing Women's Conference was designed to review and appraise the advancement of women since the Third World Conference on Women, held in Nairobi in 1985. The resulting

Platform for Action consolidated and reaffirmed earlier gains from the recent international meetings. The focus of this document was on rights rather than needs, and the universality of women's human rights was affirmed. It called for much stronger action on gender-based violence against women, and asserted that rape during conflicts is a war crime. The right of women to control their own fertility and to be free from coercion in their sexuality was reaffirmed. The document stressed the right of women to full and equal access to economic resources, including the right to the inheritance and ownership of land, and called for the measurement and valuing of women's unremunerated work. The rights of girls, including their right to freedom from such traditional practices as excision, were the subject of an entire section. A more extensive summary of the Platform of Action follows.

Summary of the Platform of Action from the Fourth World Conference on Women

The critical areas of concern identified by this document are the following:

- the persistent and increasing burden of poverty on women;
- inequalities and inadequacies in, and unequal access to, education, training and health care;
- violence against women;
- effects of armed and other conflicts on women;
- inequality in economic structures and policies, as well as in productive activities, and access to resources;
- inequality between men and women in the sharing of power and decision-making at all levels;
- lack of respect for and inadequate promotion and protection of the human rights of women;
- stereotyping of women in all communication systems;
- inequality in women's access to and participation in all communication systems;

- gender inequalities in the management of resources, including natural resources;
- the persistent violation of the rights of and discrimination against girls.

Section C on 'Women and health' in Chapter IV has five strategic objectives: (a) to increase women's access throughout life to appropriate, affordable and quality health care, as well as to information and other related services; (b) to strengthen preventative programmes that promote women's health; (c) to undertake gender-sensitive initiatives that address sexually transmitted diseases (STD), HIV/AIDS and sexual- and reproductive-health issues; (d) to promote research and disseminate information on women's health; and (e) to increase resources and monitor follow-up for women's health.

Paragraph 95 deplores the discrimination against girls that often results from a preference for sons over daughters. It also exposes the grave health risks caused by societal conditions that subject girls to harmful practices such as circumcision and that also force them into early marriage, pregnancy and child-bearing. The right of women and men to be informed and have access to family-planning methods, as well as to other legal methods for the regulation of fertility, is recognized as an instance of the right of women to control their sexuality.

The document as a whole calls for the empowerment of women in all areas. There is recognition of the complex relationship between sexual exploitation, women's subordination and their being infected by sexually transmitted diseases, including the HIV virus that leads to AIDS. Recommendations call for legislation against socio-cultural practices that contribute to women's susceptibility to HIV and STDs.

Governments are called on to allow women access to affordable social-security systems in order to ensure an equal footing with men. They are asked to take measures to eliminate harmful, unnecessary or coercive medical interventions and inappropriate medication for women.

Health is mentioned in almost every section, with links made to poverty, education, violence, human rights, the environment and girls. Governments are called on to review legislation containing punitive measures against women who have undergone illegal abortion.

The key recommendations call for the following actions in the following areas:

- women and poverty: encourage research into the ways poverty increasingly affects women; establish adequate safety nets and social-security systems to place women on an equal footing with men;
- education and training of women: promote equality of access to education and training of all kinds, a gender-sensitive curriculum and women's full involvement in the planning and implementing of education policy;
- women and health: develop gender-sensitive health information; sponsor research on women's health problems; promote recognition of women's sexual and reproductive rights;
- violence against women: encourage research into the causes and consequences of violence; establish steps to prevent and eliminate violence against women;
- women and armed conflict: encourage women's contribution towards a culture of peace; provide protection to refugee and displaced women;
- women and the economy: promote harmonization of work and family responsibilities for men and women; eliminate job segregation and all forms of employment discrimination; provide high-quality, flexible and affordable child care; seek to measure unpaid work;
- women in power and decision-making: introduce equal access to and full participation in decision-making; increase women's capacity to participate in decision-making and leadership;
- institutional mechanisms for the advancement of women: strengthen and provide adequate resources in official bodies

responsible for promoting equality; evaluate policies and programmes for their impact on women and men respectively;
- human rights of women: ensure full implementation of all human-rights agreements; promote equality and non-discrimination in practice and in law; ensure that women know their rights and how to enforce them;
- women and the media: increase women's participation in and access to the media; promote a balanced picture of women in the media;
- women and the environment: involve women in decision-making at all levels; ensure that development programmes respond to women's needs; encourage research on the impact of environmental policies on women;
- girls: promote equality for girls in all aspects of their lives; eliminate negative cultural attitudes and practices against girls; strengthen the role of the family in improving the status of the girls.

Recent commonwealth initiatives[1]

The general spirit of these United Nations conferences was echoed at the Eleventh Commonwealth Health Ministers Meeting (1995) on the topic of 'Women and Health'. This meeting recognized that many countries had innovative projects addressing issues of concern to women and discussed ways of supporting them. One priority recommendation resulting from the meeting called for the collection and dissemination of models of good practice in health concerns related specifically to women's issues. The recommendation also called for the Commonwealth Secretariat to provide annual Commonwealth Awards for

1. The following section is derived from *Models of Good Practice Relevant to Women and Health*, Commonwealth Secretariat, London, 1997.

Excellence on the basis of the findings. In so doing, this initiative aimed to set standards for good practice in women's health and to profile, and thus assist, the development of approaches that address the most serious disadvantages women experience with regards to health concerns.

Principles of Good Practice in Women's Health

The Principles of Good practice in Women's Health were developed from the growing literature on women's health and from the platforms of action of the three recent United Nations conferences on women discussed above. The principles were formulated in general terms so that they could be applied to research, training, evaluation, policy and intervention strategies. As a result, it was recognized that not all projects addressing women's health would fulfil all of the criteria listed. Revised after input from a panel of international experts from a range of countries, and presented at several international conferences for further comment, the framework was then piloted by the Secretariat in Zimbabwe and South Africa. A brief overview of the principles follows.

Scope

1. Women's health concerns extend over the life cycle and are not limited to reproductive problems.
2. Women's health problems include, but are not limited to, conditions, diseases or disorders which are specific to women, or that occur more commonly in women or have different risk factors or duration when suffered by women rather than men.
3. Health must be considered in broad terms, both positively and negatively; dimensions of health include the physical, mental, social and spiritual.

Determinants

4. Women's health is directly affected by a range of socio-cultural, physical and psychological factors.
5. Women have gender roles and responsibilities that directly affect their access to and control of the resources necessary to protect their health; these resources are 'external' (economic, political, information, education, environment) as well as 'internal' (self-esteem, initiative).
6. Women are diverse in age, class, race, ethnicity, religion, functional capacity, sexual orientation and social circumstances. These factors may lead to inequities that adversely affect their health.

Community participation

7. Priority should be given to issues that have been identified as important by women themselves. Particular attention should be paid to those issues raised by women who are subject to inequities in their society.
8. Women from the target community should be involved in the planning, implementation and evaluation of projects involving their health.
9. Knowledge arising from projects must be accessible to all women, and particularly to women in the target community. This implies that information must be provided in a fashion that is intelligible to these particular women.

Methods

10. In order to address the issues affecting women's health, a general, interdisciplinary, gender-sensitive approach is needed, involving the knowledge and methods of social and health scientists, as well as those of other disciplines.

11. Intersectoral approaches are needed to address the social factors affecting women's health and life expectations. These may involve the collaboration of various governmental departments, non-governmental organizations, community groups and private associations or foundations.
12. Knowledge originating from projects should inform and influence government projects, policies, plans, legislation and research. It should also influence health-care workers.
13. Where possible, there should be a sharing of skills within regions.

Commonwealth Secretariat Awards

In the first round of awards held by the Commonwealth Secretariat, over ninety applications were received from a wide range of countries of diverse cultural and economic status, varying from island atoll states to western industrialized nations. The applications covered important issues affecting women's health throughout the lifecycle. The sheer diversity of issues, as well as approaches used to address them, bore witness to this inherent cultural variety, but nevertheless incorporated most of the principles of good practice.

The first set of these principles emphasizes the need for a general scope. Thus, the School Nutrition Programme, submitted by the Aga Khan University in Pakistan, aimed to improve children's health as well as their ability to retain knowledge taught at school, mandating the involvement of girls. The links between education and health have been widely acknowledged (World Bank, 1993), and although enrolments are growing among girls in developing countries, more girls than boys are either still denied access to schooling or drop out at an earlier age (UNDP, 1995). Projects that focused on reproductive problems also illustrated the benefits of this broader scope. The Women's Centre of Jamaica Foundation dealt simultaneously with the problems of teenage pregnancy, interrupted education and low-employment potential. The Small

Families by Choice project in India integrated family planning and sexual- and reproductive-health services with programmes aimed at developing women's skills and literacy.

Such recognition of the multiple determinants of health is also well illustrated by a programme received from Zimbabwe, entitled Promoting Women's Health and Development through Functional Literacy Generating Activities and Intersectoral Action. Emphasizing the diversity of women's experiences that can exist even within one country or region, this project chose the most disadvantaged and economically vulnerable women as the focus.

A theme common to all the projects is that of community participation. Consultations or meetings with groups of women were used to identify the issues that needed to be included in health policies and in the distribution of information concerning health. Providing accessible information has meant the use of innovative strategies that differ from country to country, depending upon the country's stage of development and the educational level of the women.

All the projects have been interdisciplinary and intersectoral. Some of them have been shared between countries from their inception. One major aim of this Commonwealth programme in documenting such a variety of projects is to facilitate skill-sharing and to promote the application of technological advances to women's health.

Examples of programmes embodying the principles

School nutrition programme, The Aga Khan University, Karachi, Pakistan

The School Nutrition Programme (SNP) is one of the eleven components of the Sindh Primary Education Development Programme (SPEDP), a part of the Sindh Education Department of the

Government of Sindh. The SPEDP is funded by the World Bank, but SNP, along with two other components, is supported by money made available to the World Bank by the Norwegian Agency for Development Cooperation (NORAD). The two other NORAD-supported components are free: textbooks for girls in primary schools and scholarships for girls in secondary schools. Together, the three NORAD-supported components form a whole within SPEDP. The SNP will attempt to promote, among parents and teachers, the health and education needs of children. It thereby hopes to increase enrolment and attendance in primary schools, especially among girls. The provision of free textbooks for girls is designed to encourage their completion of primary school. Once girls have completed their primary education, special scholarships for girls will attract them into secondary schools and help them to complete their secondary education.

The SNP is a pilot project and not a pre-conceived package to be implemented according to a preconceived time frame. This flexibility is used to steer through obstacles and work with the advantages found in village communities in Pakistan. The project is attempting to find out what is feasible within the context of a variety of socio-cultural norms in a society struggling through transition. It is also trying to find new ways to solve problems that would help people become confident enough to take responsibility for planning and implementing their own actions. The actual programme will provide funding, support and training to help rural villages run a feeding programme within a nearby school. The feeding programme is more than just a food incentive to attract children to school. It is also a tool for the promotion of higher standards of nutrition, especially where children are concerned, and for the promotion of the recognition of schools by the villagers.

Women's Centre of Jamaica Foundation, Kingston

The main focus of the Women's Centre of Jamaica Foundation (WCJF) is the operation of its Programme for Adolescent

Mothers. This programme deals with the problems associated with teenage pregnancy, namely, interrupted education and low employment potential among young women. Girls who become pregnant while still at school are referred to their nearest Women's Centre by counsellors, teachers, prenatal clinics and government and private agencies. An integrated Day Programme is offered at each Centre and comprises academic instruction, counselling and training in areas such as sewing, chicken and fish farming, cosmetology and doll- and pillow-making. Overall, the programme is designed to make young mothers more independent. The Foundation now operates seven main centres and fifteen outreach stations. Other services offered by the Foundation include the following: a walk-in counselling service for women; counselling for the fathers of babies and for the parents of teenage mothers; training for both males and females in the 17–25 age group in the Kingston area and in neighbouring rural parishes; a homework project for youngsters aged 9–12, which includes academic instruction, family-life education and counselling; a counselling programme for Grade 7 students in secondary and all-age schools; a confidential counselling service for children of any age; group peer-counselling sessions; and day-care facilities for the babies of working mothers.

Small Families by Choice, International Planned Parenthood Federation, South East Asia Regional Bureau, Madhaya Pradesh, India

In northern India the practice of family planning is rare. The four states of Bihar, Rajasthan, Madhaya Pradesh and Uttar Pradesh are home to 40 per cent of the country's population and experience higher than average birth and death rates. Although there is a strongly felt need for family planning in this region – one half of Indian couples who are not using family-planning methods but who have indicated they do not want more children live in these states – family planning remains 5 per cent to 20 per cent lower than the national average. The overall situation concerning the development of the region is also difficult. The status of women remains very low. Early marriage and child-bearing are the norm. Moreover, in a cultural context where sons are preferred to daughters, most women and girls are illiterate, have little potential to earn an income and suffer from poor health and nutrition. The family-planning infrastructure is inadequate, with insufficient outreach services, a lack of commitment to training, and the use of counter-productive incentives and poor motivation strategies.

The Small Families by Choice project has been designed to address such needs through innovative, community-based strategies in three of the ninety worst performing districts of India: Bhopal, Vidisha and Sagar, in the State of Madhaya Pradesh, where birth rates are as high as 30 per 1,000 inhabitants. The project aims to encourage the norm of the small family among 4.3 million people in 3,910 villages, while demonstrating that such strategies can be used to reach and improve the national average in birth rate and contraceptive use. The integration of family planning and sexual and reproductive health with development and health infrastructure is a major principle of the project, which views overall development as a significant stimulus for family planning.

The $10.8 million, five-year project began in January 1995. It was the first and largest project accepted for funding by the IPPF's Vision 2000 Fund. The Vision 2000 Fund offers funding on a competitive basis for innovative projects designed in accordance with IPPF's Strategic Plan. The project demonstrates effective, innovative strategies with an integrated outlook. The project also focuses on raising the age at which girls marry, reducing maternal mortality and improving access to safe delivery, reproductive-health choices and family-planning services. Designed to develop women's skills and their potential to earn money, these initiatives have been strengthened by education and training. In addition to family-planning indicators, targets have also been set to meet the following goals: the immunization of children; the creation of local voluntary groups; the establishment of community-based distribution outlets for contraceptives; increased co-ordination with local NGOs and local government; the development of in- and out-of-school youth training programmes; and the creation of women's clubs and women's skill and literacy centres.

Promoting Women's Health and Development through Functional Literacy Generating Activities and Intersectoral Action, WHO Focal Point, Harare, Zimbabwe

This project's main aim was to focus attention on vulnerable groups of women and to show how interventions could be implemented by government services. Women were chosen for their illiteracy, their extreme poverty and for their having lost a child from a preventable disease. The women had also to be within the reproductive years with at least one child under five years old. The project addressed the following areas: the state of women's health; their ability to cope in the world; the development of functional literacy skills; and income-generating activities, designed to give women additional resources, develop their confidence and assist in the achievement of good living habits.

After its inception in 1987, with considerable government input and with seed money from the World Health Organization, the project has succeeded in several areas. Sanitary facilities, immunization and better reproductive health care have all been provided, and the women's awareness and appreciation of health issues have increased. Though frequent droughts have stalled the process, safe water has become more widely available, and more women are undertaking good hygienic practices such as boiling water. Some functional literacy classes have been well attended. But attendance patterns are strongly correlated both with economic activity and with the increased confidence and ability to communicate exhibited by women able to earn an income.

The project's greatest difficulties have involved establishing viable economic ventures. Further help is still required in this area. Some viable activities have been set up, such as market gardening and raising rabbits; others, such as a tie-and-dye project, a feed lot for cattle and poultry, and a grinding machine for maize and small grain, require further work.

The project is currently limited to certain very problematic regions, but it is planned to expand to other rural communities. In so doing, it is hoped that the project will create a common ground of understanding about the needs of vulnerable women, and thus that it will be instrumental in advocating changes in policy and the allocation of resources.

Conclusion

The work of the recent United Nations conferences and that of the Commonwealth Secretariat draw attention to the fact that, in many regions of the world, and irrespective of the wealth of the region, many women experience significant disadvantages in health concerns. Moreover, the conference recommendations and the Commonwealth support for local programmes represent exemplary steps towards redressing real inequalities.

Nevertheless, the questions ultimately raised by the issues discussed above are ethical in nature, concerning autonomy, equality, distributive justice and human rights. It has been shown that women are often denied such fundamental rights as the right to control their health and the right to education. They often suffer unnecessarily from being denied the possibility of controlling their own fertility, and, sometimes as a direct consequence of this, and sometimes for other similarly abhorrent reasons, they suffer unnecessarily in childbirth. Issues such as these raise further questions about equal rights and access to care. Although the work discussed above is significant in its attempt to bring attention to these issues, and to promote the redressing of them, there is undoubtedly a need for laws to uphold the ethical principles in question. It is part of the role of bioethics to clarify these principles and, furthermore, to determine how they should be incorporated into the existing moral codes of the respective countries.

Women's health across the life cycle

Penny Kane*

It is commonly held that women around the world tend to live longer than men and that this difference in longevity increases as life expectancy itself increases. Although this claim may be true, it results in the generalization that women enjoy better health than men. In fact, at all ages and in all societies, women's health experiences differ from those of men. Indeed, women have generally been expected to experience different patterns of mortality than men. But up until now, in many countries, and especially on the African continent, general information concerning mortality has been highly unreliable. However, a recent study undertaken by the World Bank and the World Health Organization presents results that allow us to reconsider this generalization about women's health. This recent work, entitled *The Global Burden of Disease and Injury* (Murray and Lopez, 1996), represents probably the most ambitious and comprehensive attempt to obtain reasonable estimates of levels of health and mortality in each of the regions of the world. Except where otherwise

* Consultant on health, population and family planning; University of Melbourne (Australia).

indicated, the present chapter makes use of the data assembled by this study.

Recent estimates of levels of health and mortality

Due to a low average age and a high childhood-mortality rate, 98 per cent of all deaths under the age of 15 years occur in the developing world. And there is considerable diversity even within developing regions. For example, more than half of all deaths in sub-Saharan Africa occur between the moment of birth and four years, compared with 11 per cent in China for the same period. Moreover, in sub-Saharan Africa more than half of these deaths are attributable to infectious and parasitic diseases, although respiratory diseases are also common causes of death. In China, on the other hand, deaths in early childhood are more likely to result from perinatal conditions and respiratory infections. In order to explain such a diversity in levels of early childhood mortality, researchers cite such significant factors as differential access to clean water, immunization and education, as well as different levels of household income (World Bank, 1993).

In some countries of Asia, female foetuses are noticeably less likely to survive through pregnancy than their male counterparts. And if they are eventually born, girls in these same countries are also less likely to survive through early infancy. The Republic of Korea currently reports the highest sex ratio at birth, with 115 males for every 100 females. Other countries with high and rising sex ratios at birth include China, Taiwan and Hong Kong SAR (Park and Cho, 1995). It is also well known that the Indian subcontinent has very skewed sex ratios, although no national surveys have been carried out to provide accurate statistics. Of course, some of the 'missing' female babies may simply be unregistered; but others have surely been the victims of sex-selective abortion, neglect or even infanticide. UNDP (1994) estimates that the difference between actual and expected survival rates of female babies world-wide implies more than a million excess female deaths annually. Even if more of those girls do survive than this estimate suggests, they are likely to live only as second-class citizens, with reduced chances of education and health care compared to their brothers.

Between the ages of 5 and 44, women in most developing countries are more likely than men to die. First of all, they are more vulnerable than men to communicable diseases. And secondly, during their reproductive years, women are prone to fatal dangers experienced during pregnancy and the act of giving birth. By contrast, in the established market economies, a higher female vulnerability to communicable diseases disappears after age 14, a change due largely to the impact of the HIV epidemic among adult males.

Tuberculosis is the leading cause of death for women between the ages of 15 and 44, accounting for about one-fifth of all deaths. About a further 7 per cent of female deaths worldwide are due to suicide, with particularly high rates found among women in rural China. Among Indian women, death by fire accounts for a quarter of all deaths from any injury. It is believed that many of these cases are actually homicides.

Some 450,000 women are estimated to die annually from causes linked to reproduction. Poor health and nutrition during pregnancy is a contributing factor. For example, high proportions of women in developing countries suffer from anaemia, and frequent childbearing results in a maternal depletion syndrome. About 186,000 of these deaths occur in sub-Saharan Africa and a further 115,000 in India. In the least-developed countries, health care during pregnancy is minimal, and more than two out of every three women give birth without the help of any trained health personnel (UNDP, 1994). Sequelae of induced abortion are believed to account for some 13 per cent (around 61,000) of these maternal deaths. The recent round of Demographic and Health Surveys found that in developing countries outside Africa at least half of those married women surveyed reported that they did not want to have any more children. In Africa itself almost half of the married women surveyed either wanted to postpone their next birth or to stop having children (Robey et al., 1992).

Measuring ill-health

Although death may be used as the most easily identified measure of ill-health, it is nevertheless an inadequate measure. The World Bank/WHO study has developed an index of the total burden of disease, which attempts to measure the impact of disability as well as of premature death. This index is described as the 'disability-adjusted life year', or 'DALY'. One DALY is equivalent to one lost year of healthy life (Table 1).

The disease pattern among children, as expressed by the DALY, does not vary much according to gender, but in the adult years there is some divergence. Among the ten leading diseases between the ages of 15 and 44, depression has the greatest effect on the DALY for both men and women in developing countries and for women in the developed world. Iron-deficiency anaemia, which as a cause of death was comparatively insignificant, now contributes significantly to the total burden of disease in developing countries, especially among women. Three separate maternal conditions appear in the ten leading causes of disease-burden for women in developing countries: if combined into a single group of 'reproductive causes' they would hold second place on the list, increasing the DALY factor by 9.1 per cent.

Also extremely striking is the extent to which mental illness figures in the causes of disease-burden. Depression, in particular, is a major contributor to health impairment in both developed and developing countries. Self-inflicted injuries, too, are prominent among the causes of disease-burden; in developing countries women are more likely to inflict injuries on themselves, as opposed to men in the developed regions.

Table 1. Ten leading factors influencing 'disability-adjusted life years', or DALYs, ages 15–44 years (1990)

	Males Disease/injury	Per cent DALY	Females Disease/injury	Per cent DALY
Developed countries				
1	Alcohol use	12.7	Unipolar major depression	19.8
2	Traffic accidents	11.2	Schizophrenia	5.9
3	Unipolar major depression	7.2	Traffic accidents	4.6
4	Self-inflicted injury	5.7	Bipolar disorder	4.4
5	Schizophrenia	4.2	Obsessive-compulsive disorders	3.8
6	Drug use	3.8	Alcohol use	3.2
7	Violence	3.3	Osteoarthritis	3.2
8	Ischaemic heart disease	3.1	Chlamydia	3.0
9	Bipolar disorder	3.1	Self-inflicted injuries	2.3
10	HIV	2.4	Rheumatoid arthritis	2.2
Developing countries				
1	Unipolar major depression	7.0	Unipolar major depression	12.8
2	Traffic accidents	6.3	Tuberculosis	4.9
3	Tuberculosis	6.0	Iron-deficiency anaemia	4.1
4	Violence	5.5	Self-inflicted injuries	3.7
5	Alcohol use	4.6	Obstructed labour	3.4
6	War	4.2	Chlamydia	2.0
7	Bipolar disorder	3.1	Bipolar disorder	3.0
8	Self-inflicted injuries	3.0	Maternal sepsis	2.9
9	Schizophrenia	2.8	War	2.8
10	Iron-deficiency anaemia	2.8	Abortion	2.8

Source: C. J. L. Murray and A. D. Lopez, *The Global Burden of Disease and Injury*, Cambridge, Mass., Harvard University Press, 1996.

Sexually transmitted diseases (STDs), taken as a group (excluding HIV), create a much greater disease burden for women than for men in all regions, although the levels of infection and the amount of excess disease burden vary widely. It is noticeable that chlamydia is amongst the ten leading causes of DALYs for women in both developed and developing countries. Sterility is often one of the consequences of STDs, and this may lead to the woman's being divorced and disgraced, which in turn may result in a lack of economic support for her.

HIV, however, has a greater impact on men except in sub-Saharan Africa. But when a man dies of an HIV-related illness, it may be the case that he leaves behind him an HIV-infected wife, who is alone to support their children, who may themselves be HIV-positive. And in countries where HIV is predominantly a male problem, women who are infected are likely to suffer delays in diagnosis and to experience less appropriate support than men.

What women, and even men, can expect from their health in old age remains ambiguous. In every region of the world, female life expectancy now exceeds that of males. However, in four regions – established market economies, Latin America and the Caribbean, former socialist economies and sub-Saharan Africa – men not only live shorter lives, but spend a larger proportion of those lives with disabilities. In the other four regions addressed by *The Global Burden of Disease and Injury* study – India, China, other Asia and islands and the Middle Eastern Crescent – it is women who spend a higher proportion of their lives disabled.

For both women and men aged 60 and over, the greatest contributors to the burden of disease are cardiovascular diseases, cancers and respiratory diseases. In most countries, women's greater life expectancy is likely to involve widowhood and reductions in income and security. Women's increasing participation in work outside the home reduces the amount of old-age care available from daughters or daughters-in-law. Other factors – including migration and increases in some countries in the proportion of displaced rural workers (Dharmalingam, 1994) – may also reduce the support available for the elderly.

Conclusion

Although *The Global Burden of Disease and Injury* study represents one of the most comprehensive attempts to compile trustworthy information on health and mortality worldwide, perhaps inevitably some concerns about women's health have not been addressed. In cases where an issue has only recently begun to be understood as a health problem for women or where it has been given little weight by the dominant medical culture, its significance has been overlooked. Violence against women, for instance, which occurs mainly in the home, and only a small fraction of which ever results in a hospital incident, is underestimated as a cause of both physical and mental illness. Another example of such an issue is female circumcision. It is estimated that this practice affects, in some form or another, between 85 and 114 million women (UNDP, 1994). Women may suffer long-term health complications as a result of the repeated cutting and stitching the practice sometimes involves, and they may also experience adverse effects on sexuality and mental well-being (see the following chapter, Case Studies). Similarly, menstrual problems, especially among women in developing countries, have received little attention from researchers and are seldom given much weight in discussions of reproductive health or psychological well-being. This is all the more surprising in light of the fact that, in two WHO multinational studies, dysmenorrhoea was a complaint of a quarter to three-fifths of the sampled women (Omran and Standley, 1976, 1981). Moreover it can be stated that data on women's occupational health remain narrowly focused, scant and often of very poor quality (Kane, 1997). There is still much we do not know about women's health; perhaps the first step in expanding that knowledge will be a recognition that women's status has defined and curtailed the subject itself.

Infants, children and adolescents

Lucile Newman*

All human beings are born free and equal in dignity
and rights. They are endowed with reason and conscience
and should act towards one another in a spirit of brotherhood.
(*Universal Declaration of Human Rights*, Article 1)

The *Universal Declaration of Human Rights* affirms a set of rights common to all. Nevertheless, this affirmation is made in a world characterized by profound cultural variation, vast economic diversity and deep gender disparity. Indeed, contrary to the affirmation of the *Universal Declaration of Human Rights*, in most societies girls and women do not share human rights equally with boys and men. In every society, no matter how economically rich, we still encounter social institutions and systems of

* Department of Community Health, Brown University (United States).
The author acknowledges support from the following people: Anke A. Ehrhardt, Lina Fruzzetti, Patricia Symonds, Erin Roland, Kyle Bernstein, Kathy Kulkarni, Antoine McCoy, Lucia Duncan, Monica Garg, Emily Grossman, Keerthi Gogineni, Kate Mason, Robin Proctor, Rachel Reichlin, Margarita Ramos and Brijen Shah.

thought that promote gender differentiation and, thus, inequality. Moreover, such inequality begins almost universally for girls in infancy, and it continues throughout childhood and adolescence.

In *A Theory of Justice*, Rawls argues that 'justice is the first virtue of social institutions, as truth is of systems of thought'. Concerning the issues of human rights and women's health, it is our task to examine these social institutions in order to see where they lack their 'first virtue', that is, where they promote gender discrimination, even poverty, and have such a strong impact on women's formative years. To this end, the present chapter will present an examination of these years in a culturally diverse context, and it will attempt to show how social institutions influence the development of males and females unequally. As such, the present chapter aspires to throw some light on ethical, bioethical and human-rights concerns in the early years of women's lives.

The important issues present throughout this general period are difficult to separate, for taken together, they define the very quality of early life. However, since the social institutions that influence this period the most are not always the same, three distinct stages will be examined: infancy (0–5 years), childhood (6–11 years) and adolescence (12–17 years). During infancy, survival, nutrition and health care are of particular concern to females. In childhood, the most important issues of human rights and equity for girls are health, nutrition and education; as girls begin to define career options, where this is possible, work also takes on significance. In adolescence, these issues encompass education, health, a developing sexuality and the wide variety of life experiences necessary to facilitate entry into and adaptation to the working world.

Infancy

Ensure to the maximum extent possible the survival and development of the child
(Convention on the Rights of the Child, Article 6)

While survival and development are both issues of human rights, in both the developed and the developing world, infancy is the period of life most subject to discrimination. The most important issues during this time have to do with survival, nutrition and health care.

Survival

Infant-mortality rates provide some of the most convincing indicators of the health risks suffered by children in the developing world. Indeed, the mortality rate of children under five is higher here than in industrialized societies. Infant-mortality rates vary from an average of around 11 deaths per 1,000 live births in developing countries to 109 deaths per 1,000 live births in the least-developed countries. Where infant and early childhood mortality are high, total fertility rates must also be high to ensure the survival of the species. Fertility rates of 5.5 in the least-developed countries contrast sharply with rates of 1.7 in the developed regions (United Nations, 1996). In general about 105 males are born for every 100 females, but mortality is usually higher for male infants, thus re-establishing the expected gender ratio (Visaria and Visaria, 1995). However, in many developing countries female-mortality rates are higher, partly due to less care, partly due to female infanticide.

Modern technological interventions enable gender discrimination to occur even before birth through prenatal amniocentesis, often conducted for the sole purpose of determining gender, and through the abortion of female foetuses (See Case Study, p. 50, 'Sex Selection in India'). This introduces a bioethical question pertaining to the use of modern technology to limit female foetal development.

The progress of medical technology has made it possible to keep some infants alive, when in fact their chances of survival are very small, leaving parents unable to cope and in debt financially. As a reflection of international concerns, present discussions at medical meetings and in journals have called such issues 'unbridled' medical intervention (Silverman, 1991). Kittrie (1974) has referred to the prevailing practice, common in developed countries, of saving all lives with 'extreme technological interventions' as 'the dominance of the savers'. In neonatal intensive care units for very low birth-weight infants, variations in procedure also reflect cross-cultural values of professionalism.

Medical hegemony functions in other ways, too. At present, infertility is increasing and compromises the chances of an ever-greater number of couples to have children. In response to this trend, there has been a development of new reproductive technologies, including *in vitro* fertilization (IVF) and gamete intra-Fallopian transfer (GIFT). The technique of IVF relies on the combination of the sperm and ovum in a laboratory dish and the subsequent transfer of the fertilized embryo to the woman's uterus. GIFT is a similar technique, except that the fertilized egg is placed in the Fallopian tubes. In order to increase the possibility of successful gestation, multiple ova are often implanted. While some may not survive, others do, and this often results in multiple gestation. A world survey in 1987 showed that 24 per cent of pregnancies resulting from IVF or GIFT were multiple pregnancies. Many of the risks associated with these two techniques are due to multiple pregnancies, such as an increase in the chance of ectopic pregnancies and miscarriage. Furthermore, in recent years we have witnessed a rise in the number of low birth-weight infants. Other problems more likely to occur are pre-eclampsia, premature delivery and neonatal mortality.

Questions concerning such interventions must take into account not only the possible effects on the woman's overall health, but also the well-being of the surviving children in terms of normal development and learning capacity.

There are other ethical issues as well. A woman may have IVF treatment with eggs donated by another woman, or there may be surrogate motherhood, where a woman gestates the embryo of another couple. The issue of payment, either for the donation of eggs or for acting as a surrogate mother, is controversial. And surrogate motherhood has the added risk that the surrogate mother may change her mind about giving up the baby or may not act responsibly during her pregnancy.

The variations in international attitudes concerning the acceptability of reproductive technologies result partly from economic factors. Recourse to such technologies is often prohibitively expensive. A pregnancy following infertility treatment costs four to five times as much as a normal pregnancy. Indeed, such bioethical and human-rights issues as those mentioned here often arise only in developed countries, and even then only in a small subset of economically advantaged individuals within those countries.

Nutrition

A bioethical issue arises in the form of the differential care given to male and female infants, including in many countries such practices as longer breast-feeding for male infants. In patrilineal societies males are considered 'members of the continuing family', while females will marry out of the family. In addition, a dowry requirement for females provides an economic incentive to limit the number of girls in the family. In South Asia, consistently higher death rates for females, from infancy to adulthood, have been attributed to a preferential treatment of sons: boys are fed first, taken earlier for health care when ill and required to attend school.

Health care

Preventing problems such as birth trauma or neonatal tetanus requires safe practices during delivery. However, the differences in health care and delivery practices between rural and urban areas result in differences in mortality. Environmentally borne diseases such as pneumonia and diarrhoea may be avoided by consistent breast-feeding. But information and education are also necessary, and these are not always available to rural populations. In India, for example, the mortality rate of infants in urban areas is always less than that in rural areas. According to a study conducted in 1993, the mortality rate of infants in urban areas was 45 per 1,000 births, compared with 82 per 1,000 births in rural areas (Visaria and Visaria, 1995). The same study addressed the issues of literacy and found that, according to a census conducted in 1991, only 39 per cent of Indian females were literate, compared with 64 per cent of their male counterparts.

Excess mortality among girls between 1–4 years occurs where female infants receive less care, fewer nutrients and less attention from parents than boys. In poor families, neglect means that girls are more likely than boys to suffer from infectious disease and are more likely to die during infancy and early childhood. Infants and very young children of both sexes are most subject to infectious disease, and the mortality rates reflect this. Immunizations are important at this time. International agencies have attempted to immunize as many children as possible, but the problem is immense.

Despite a biological advantage that favours girls, especially during infancy – girls are more likely to survive if equal care is provided – gender disparity begins early. One pervasive pattern that becomes apparent worldwide is a disadvantage for girls just to survive. In patrilineal families and families where male children are prefered, the risk factors stretch from one generation to the next. Because of the way value systems are seemingly inevitably perpetuated, a mother who had already suffered gender preference, perhaps in the form of malnutrition during childhood and limited access to education, will probably cause her female children to endure a similar fate, perhaps in the form of shorter breast-feeding, less secondary feeding and less preventive health care. The results of such discrimination include inadequate nutrition, poor overall health and, ultimately, death (Kurz and Prather, 1995).

Childhood

Ensure that the child has access to information [. . .]
aimed at the promotion of his or her social,
spiritual and moral well-being and physical and mental health.
(*Convention on the Rights of the Child,* Article 17)

Childhood is characterized by beginnings: the beginning of education, the beginning of independence and the first encounters with others in play. During this period, the most important issues of human rights and equity for girls are health care, education, nutrition and an initial experience of role playing. Taken together, these factors define the quality of life of girls between 6–11 years.

Health care

Most health care for children is provided in the home, except in areas where government programmes of preventive health care include inoculations against diphtheria, whooping cough (pertussis) and tetanus (DPT), vaccinations against polio and tuberculosis and, in some areas, vision and hearing tests.

Education

Recent international conferences, including the International Conference on Population and Development (Cairo, 1994), the World Summit for Social Development (Copenhagen, 1995) and the Fourth World Conference on Women (Beijing, 1995), all promoted investments in education for the purpose of achieving gender equality in developing countries, and particularly in those countries where there are low levels of literacy and especially low levels of education available for girls (Table 1). In the developing world, the ratio of literate women to literate men is about 71 : 100; in the least-developed countries, the ratio falls to 54 : 100. While almost as many girls as boys now enter primary school in developing countries, only 72 girls for every 100 boys progress to secondary education, and twice as many young men as women go on to tertiary education (UNDP, 1994).

Table 1. Female and male literacy in four French-speaking countries of Africa, 1990

	Female (per cent)	Male (per cent)
Burkina Faso	9	28
Mali	24	41
Niger	17	40
Senegal	28	52

Source: World Resources Institute, 1992–1993.

Nutrition

The connection between poverty, malnutrition and cognitive development has become specific in relation to micronutrients such as iodine and iron. The general concern has been with iodine deficiency and its effects on children's levels of intelligence. Indeed, iron-deficiency anaemia is considered to be the most prevalent nutritional problem in the world today. Early care as well as attention to nutrition are important factors in ensuring regular brain development in infants and children. Gender discrimination in South Asia, where traditions stipulate that girls should be fed last, contributes to the highest rates of child malnutrition in the world (Ramalingaswami, Jonsson and Rohde, 1996).

Work

In South Asia the dowry requirement for females creates difficulties in poor families. Girls may be abandoned; or 'loaned' or sold to work in factories or fields to earn money for another sister's dowry. Young girls may be found at train stations or on the street, others outside orphanages. Although they are cared for and trained, they are without name, identity or family ties, and they may not be marriageable except to residents of another state, or outside India, if at all. Fruzzetti notes that 'the most exploited children belong to the most marginalized segments of society, including ethnic minorities and disadvantaged groups' (1997).

The exploitation of children has occurred in many different contexts in many countries, including the factories of the developed world and the fields of agricultural societies (United Nations, 1997). Children are often forced to work in hazardous conditions such as those found in fields where pesticides are used and in factories where unusually long working-hours are imposed. In many parts of the world, including developed countries, immigrants, legal or illegal, and itinerant populations have children doing agricultural and factory work under these very dangerous conditions. In addition, poor children have been used for begging and sexual exploitation in most countries of the world; sometimes they have even been transported to other countries for just these purposes. In addition to grave effects on general health, enforced labour during early childhood results in a lack of education and thus in long-term poverty (UNICEF, 1997).

Children, some unaccompanied by families, have been refugees from wars in many parts of the world. Accounts of the lives of Guatemalan and Salvadorian children in Belize indicate pervasive health problems such as malnutrition, dehydration, diarrhoea and respiratory illnesses (Miller, 1996; Moss, 1992). Dowell (1995) has stated that among Rwandan children living in camps in the Democratic Republic of the Congo, 40 per cent of deaths were caused by dehydration due to acute diarrhoea.

Adolescence

Education shall be directed to the full development
of the human personality and to the strengthening
of respect for human rights and fundamental freedoms.
It shall promote understanding, tolerance and
friendship among all nations, racial or religious groups,
and shall further the activities of the United Nations
for the maintenance of peace.
(Universal Declaration of Human Rights, Article 26 [2])

During adolescence, the transition period between childhood and adulthood, a girl faces her changing identity and her developing sexuality – in short, the beginnings of her life as a woman. During this period, a connection to and a separation from the family are important factors in a girl's developing identity. In addition, as adolescents mature, changes in physical appearance dramatize the differences between females and males. The 'gender intensification hypothesis' suggests that socialization and role expectations are most evident in adolescence (Lynch, 1991). The hypothesis further suggests that this is the period when physical appearance and social skills are emphasized for girls at a time when autonomy and achievement are stressed for boys (Lynch, 1991). The human-rights inequities experienced by girls in this period of their lives often set a tone of unequal status that will not change in adulthood.

Education

In societies where agricultural labour or participation in a family business begins early, adolescence marks the beginning of work. In societies with public secondary-school education, full-time employment may wait until education is completed. Differences in the extent of secondary education occur not only between developing and developed countries, but also between poor and wealthy populations within them (Tables 2 and 3). Differential education has lifetime implications for a women's capacity to achieve autonomy and occupy an advanced professional position. To emphasize a young women's natural reproductive capacity, ignore her individual talents and confine her to menial work is to limit her access to economic advantage and deny her human rights.

Health and sexuality

Adolescence also marks a period of sexual exploration. Because over the past few decades many countries have experienced a drop in the age at which intercourse first occurs and a dramatic increase in both sexual experimentation and sexual exploitation, there has been a consequent rise in the incidence of adolescent pregnancy and sexually transmitted diseases, including HIV infection. Some causal factors of these types of behaviour include earlier sexual maturation, a lack of pertinent general knowledge, less agreement about acceptable behaviour, declining religious influence and urbanization. Many societies have been reluctant to teach sex education to females either because of religious restrictions on what may be taught and discussed, or because of a general reluctance to consider that this aspect of life is related to health, or because opponents consider that 'sex education itself will seduce adolescents into sexual activity' (Ehrhardt, 1996).

The objectives of sex education, at home or at school, are to prepare individual societal members for healthy and satisfying experiences of sex and sexuality (Ehrhardt, 1993). Sex education has been effective in the Netherlands, Sweden, France and England. These countries have established healthy perspectives concerning sex, which they feel has caused the number of unwanted pregnancies to decrease. Community-based organizations may be important in providing information and education to young people on subjects related to maternal and child health, including high-risk behaviour in the context of the HIV virus (Simbulan, 1995). In many countries in Africa and the Middle East, attempts to control female sexuality involve female genital mutilation (see Case Study, p. 58, 'Excision').

One of the many unfortunate effects of the AIDS epidemic has been a perception that the lower a girl's age, the less likely she is to be already infected with HIV, and therefore the better she is as a 'safe' sexual partner. The result has been to negociate with families to 'sell' their daughters for transportation to the large, cosmopolitan cities of India and Thailand, as well as to other countries, where they become sex workers and, on a daily basis, are exposed to HIV and other sexually transmitted diseases (Fruzzetti, 1997; Symonds, 1995).

Body image

'Body image' and a fear of appearing overweight are problematic issues for adolescent girls in many industrialized societies. Preoccupation with appearance can cause unhealthy dieting (which sometimes leads to malnutrition) and eating disorders. Anorexia, bulimia, compulsive overeating and other eating disorders occur in situations where females, and sometimes even males, feel unwelcome in society 'as they are' and go on diets to attain what they consider as 'acceptable' appearance.

Conclusion

Deep gender disparity is encountered worldwide as early as infancy, and the experiences of childhood and adolescence often only exacerbate this division. In order to address the issue properly, we must recognize the social institutions that perpetuate inequality and how they do so at each stage of development. Only when we have imposed equilibrium in our social institutions will we restore equal treatment, and consequently justice, within them.

Table 2. A world view of education

	Primary school enrolment 1990–1995, per cent		Secondary school enrolment 1990–1995, per cent	
	Male	Female	Male	Female
Sub-Saharan Africa	58	50	26	21
Middle East / North Africa	92	82	62	49
South Asia	–	–	51	32
East Asia / Pacific	97	95	57	49
Latin America / Caribbean	86	86	45	49
CEE / CIS and Baltic States	98	96	80	82
Industrialized countries	97	97	97	99
Developing countries	86	81	51	41
Least-developed countries	56	45	21	12
World	88	84	57	49

Source: UNICEF, 1997.

Table 3. A world view of literacy, 1970 and 2000

	1970			2000		
	MF	M	F	MF	M	F
World	63	71	55	79	85	74
More developed regions and countries in transition	95	97	94	99	99	98
Less developed regions	47	59	35	73	81	66
of which						
Sub-Saharan Africa	29	39	19	61	69	54
Arab States	29	44	15	62	73	50
Latin America/Caribbean	74	78	70	88	89	88
Eastern Asia/Oceania	56	70	42	86	92	80
Southern Asia	32	45	18	56	67	44
Least developed countries	27	39	16	51	61	41

Source: UNESCO, *World Education Report 2000*.

Case study

Sex selection in India

K. Kusum*

The Constitution of India (1950) affirms the right of equality, prohibits discrimination and seeks to protect the dignity of women. The Indian Penal Code (1860) punishes abortion unless sought or performed to save the life of the woman. The Medical Termination of Pregnancy Act (1971) also prohibits abortion conducted on the grounds of knowledge about the gender of the foetus.

Nevertheless, and in spite of these legal documents, in recent years India has experienced a dramatic increase in gender-based abortions. In the mid-1970s, medical research in the field of reproductive biology aimed at developing tests to detect foetal abnormalities with the aid of amniocentesis revealed that such tests could also determine the gender of the foetus. The Indian Council for Medical Research subsequently imposed a ban on these tests. But once the technology had become known, it appeared to coincide with a certain pro-male sentiment among the population, and medical entrepreneurs went on to exploit this sentiment in order to market the technique. Their work was successful. The result is that, today, and in spite of the laws referred to above, in no other country is the misuse of medical technology to abort female foetuses through a network of clinics and centres so blatant as in India.

Consequently, India has felt the need for an effective law to curb the specific practice of sex detection followed by the abortion of a female foetus. The legal documents mentioned at the opening of this Case Study have been ineffective in curbing this

* Associate Research Professor, Indian Law Institute, New Delhi (India).

procedure. Indeed, the Penal Code is over a century old and was written when sex-detection procedures were inconceivable. The Constitution of India is a general text enshrining the principles of equality and denouncing derogatory practices, but it holds no specific weight against such modern scientific procedures. Even when the more recent Medical Termination of Pregnancy Act took force in 1971, the practice of prenatal sex-detection was still unknown. Therefore, given the sharp rise in the incidence of cases, the need for a specific law has made itself keenly felt.

Cultural bias in favour of males

The reason for the popularity of this medical technique is parental preference for male children over female children.[1] There are various socio-economic, cultural and religious factors which are responsible for this. A son is thought to provide financial and thus emotional security to parents in old age; he continues the family name and business; and he deals with the parents' funeral and last rites. A daughter, on the other hand, is an economic burden because of the dowry custom, and parents suffer constant fear of her being harassed and ill-treated by her husband or in-laws because of the dowry. In addition, in a society that places a very high value on chastity, it is considered a heavy burden to preserve the virginity of girls. Thus, sons are welcome, whereas a daughter, especially if she is not the first, is only tolerated. Some communities have resorted to female infanticide to get rid of

---

1. The obsession to choose the sex of a child is widely promulgated. For example: one full-page advertisement in a diet book has the following title: 'Boy or Girl: Choose your Child through Diet' (*Stardust*, July 1985); and we read of counselling services offered by reproductive endocrinologists designed to advise couples on the proper timing of intercourse to produce a child of the desired gender (*Saavy*, April 1985). The traditional blessing given to brides, 'May you be the mother of a hundred sons', may reflect a preference for male children in Indian society.

unwanted female babies.[2] Couples who do not wish to have daughters feel that the detection of gender followed by abortion is an acceptable practice.

Beginnings

The first clinic to perform gender-detection tests was inaugurated in Bombay in 1977, and the tests were advertised as 'humane and beneficial'. Around the same time, another clinic in Amritsar, in the State of Punjab, began a large-scale operation by placing slogans on billboards around the city, saying: 'Invest 500 rupees now, save 50,000 rupees later'. The message was clear: 500 rupees represented the cost of the test and 50,000 rupees represented the marriage expenses of a daughter. As one reporter who visited a clinic conducting these tests stated: 'Women line up waiting for their turn, the majority of them accompanied by their mothers-in-law. Perhaps the most revolting act of the doctors was to keep the foetuses of twins (both females) preserved in a jar to show to women who were hesitant about going through the test. The doctors also [used] a woman on whom they had performed the test so that she could tell other women how happy she was that she had been able to get rid of her foetuses of female twins' (*Facets*, 1982).

The business continued smoothly until a diagnostic error led to the abortion of a male foetus. This caught the attention of activists and a national debate transpired. As a result, a practice that had been known only to some people in certain areas received wide publicity. People in large numbers started using this

2. Although this practice was banned by law in 1870, reports of female infanticides from certain parts of the country are not uncommon. See, for example, 'Born to Die', *India Today*, June 1986; Usha Rai, 'Female Infanticide Thrives in South', *Indian Express*, 8 Aug. 1992; Pushpa Iyenger, 'Girls in Salem are Born to Die', *Times of India*, 30 Aug. 1992; Elisabeth B umiller, *May You be the Mother of a Hundred Sons*, Penguin Books, 1991.

'boy-girl' test. By 1985 the government had made these tests illegal. But this only resulted in the privatization and commercialization of the practice, and the procedure normally used was amniocentesis or ultra-sound. The tests were advertised on billboards, posters and handbills. As a result, the practice assumed alarming proportions as the number of clinics mushroomed in various parts of the country. According to one study, uprooted agricultural labourers and small farmers took out loans at high rates of interest to be able to pay for the test (Kulkarni, 1986).

Distressing revelations

In 1986 a group of activists launched a campaign against these tests through a group known as the Forum Against Sex Detection and Sex Pre-selection (FASDSP). This forum brought different groups together and attempted to promote public opinion against the practice. It initiated a campaign, participated in debates and seminars on television and collected information on the use of tests that led to the abortion of female foetuses. A sample survey sponsored by the state government was conducted in the city of Bombay. This survey made the following revelations:

- nearly 85 per cent of gynaecologists who took part in the Bombay survey performed amniocentesis tests with the sole purpose of sex detection;
- nearly 80 per cent of these gynaecologists performed abortions after the test if the couple so desired;
- doctors in Bombay were conducting an average of 270 tests per month;
- it was mostly women with two or three daughters who sought to be tested;
- a private clinic in a Bombay suburb that advertised its services in local trains reportedly performed 15,914 abortions between 1984 and 1985, and 99 per cent of these were preceded by the tests;
- between 1982 and 1987, the number of clinics offering the test increased from less than 10 to 248 in Bombay alone.

These revelations were echoed in results in other studies:

- Records in three hospitals in the city of Pune (Maharashtra), covering the period between June 1976 and June 1977, showed that 700 women had sought the test in a hospital, and that out of the 450 women who were informed they were carrying a daughter, 430 decided to have an abortion. All of the 250 women who were informed they were carrying a male foetus continued with the pregnancy, even those women who were warned of potential genetic disorder (Ramanamma and Bambawali, 1980).
- Between 1978 and 1983, 78,000 female foetuses were destroyed in India based on the results of amniocentesis tests (Anjali, 1987; Lingam, 1991).
- A platform against these tests, known as the *Garbhling Parikshan Virodhi Manch*, reported that 2,400 tests were conducted in Baroda city in Gujarat in 1987 and only 3 of the 30 doctors interviewed were against these tests.
- According to one report, in seven Delhi clinics, 13,000 sex-detection tests were conducted between 1987 and 1988 (Rai, 1992*a*).
- Another recent survey, conducted by the Department of Anatomy at a medical college in Jaipur in the state of Rajasthan, indicated that 3,500 female foetuses are aborted every year after the test in that city alone. The report stated that the use of widely available ultra-sound facilities for the detection of the gender of the foetus provoked the abortion of ten female foetuses daily at three medical centres in the city where the tests are carried out (UNI Report, 1994).

Legislative intervention and public reaction

In the state of Maharashtra, public outrage following these reports led to legislation to ban the tests. Despite a great deal of opposition to such a law, the state passed the Maharashtra Regulation of Prenatal Diagnostic Techniques Act (1988). However, since the law was confined to this particular state, it did not have the desired result: women sought the tests in bordering states, which had not yet passed similar laws.

The number of female foetuses aborted following sex-detection tests continued to soar, and a strong need for a central law to fight this practice was felt by some women's activist groups. Consequently, the central government set up an expert committee to look into the matter. This committee submitted its report in 1989, and in 1991 the Prenatal Diagnostic Techniques (Regulation and Prevention of Misuse) Bill was introduced. Public opinion of the document was solicited, and, as expected, the legislation received a mixed response.

Arguments for and against a legal ban

Those who are against a legal ban on sex-detection tests followed by abortion argue that such a ban would be against the interests of women. The following reasons are given: (a) it would be an infringement upon a women's freedom of choice; (b) the number of unwanted girls would rise; (c) it would lead to repeated pregnancies until a male child were born, thus affecting the health of the woman and the quality of life of the family; and (d) it would encourage clandestine practice in sub-standard medical conditions and lead to exploitation.

Proponents of this point of view also argue that such a ban would be against the demographic needs of the country. In a seminar on amniocentesis held in Bombay in 1984, a family-planning official is reported to have made the following remark: 'Our population growth has reached such an explosive situation that desperate measures are called for. So we must allow them to have the test. . . . The government's despair about population growth has found an ally in people's despair about producing daughters' (Balasubramanyan, 1986).

Those who oppose the practice argue that failure to impose a ban would allow an imbalance in the male/female ratio already apparent in the census reports to increase (Table 1). This in turn would give rise to various social problems, such as incest, rape and abduction, and would also encourage polyandry. Furthermore, the repeated cycle of 'test, abortion, new pregnancy, new test, etc.', which can continue uninterrupted until a male child is detected, would pose a threat to women's health, in some cases even to their lives, because of the repeated abortions.

Table 1. Sex ratio in India, 1901–1991

Census year	Girls per 1,000 males
1901	972
1911	964
1921	955
1931	950
1941	945
1951	946
1961	941
1971	930
1981	934
1991	929

Source: Bose, 1991.

The special law

Stringent legislation to ban sex-detection tests that are followed by the abortion of a female foetus was seen as the only solution. After the Bill was introduced in 1991 and in spite of opposition, it became an Act in 1994. Some of the salient features of the Act are the following:

- the law is all-India legislation that regulates, not bans, the use of prenatal diagnostic procedures;
- prenatal diagnostic techniques may be used only to detect gender or metabolic disorders, chromosomal abnormalities, certain genetic malformations or sex-linked disorders;
- compulsory registration of every genetic clinic, laboratory or centre with minimum standards is prescribed;

- only those women who fall in the risk category as specified by the Act can have the test performed; the conditions they must meet are the following: (1) a woman must be over 35 years old; (2) she must have a history of two or more abortions or of foetal loss; (3) she must exhibit exposure to potentially teratogenic agents, such as drugs, radiation, infection or hazardous chemicals; (4) she must have a history of mental retardation or physical deformities that are recognized as the result(s) of genetic disease; (5) she must exhibit any other conditions as may be specified by the Supervisory Board;
- advertisements of any form promoting sex detection are prohibited;
- provisions will be made for setting up various bodies to look into the issues of policy and implementation; the composition, powers and functions of these bodies are defined in the Act;
- penalties for the misuse of diagnostic procedures will be imposed; punishment will include a maximum prison-term of 3 years and a maximum fine of Rs. 10,000 in the case of registered medical practitioners, licences can be suspended;
- in the case of a woman, unless proof is given to the contrary, there is a presumption that she has been compelled by her husband or relatives to undergo the test and so he/they will be liable for abetting the offence;
- the offences under the Act are cognizable and are not subject to the terms of bail.

Conclusions

Problems and practices that have their roots in the social and cultural milieu of a society must be addressed by legislation. It is hoped that such legislation will promote a change in prevalent attitudes. In the context of India, one desired result of the legislation discussed above has been a change in people's perception of the worth of girls.

Although a ban on sex-determination tests that are followed by the abortion of a female foetus is undoubtedly necessary to bring about such attitudinal changes, other measures should also be pursued. These should aim, among other things, to ensure the dignity, safety and protection of females in and outside the home; establish stringent anti-dowry laws; and provide security for the elderly so that parents do not feel they will have to rely on protection only from their sons.

Case study

Excision

Nahid F. Toubia*

The practice of excision[1] is an initiation ritual in certain societies intended to mark a girl's acceptance of her womanhood and to define her within the female social norms of her tribe or ethnic group. Although excision is part of an ancient cultural tradition, the practice itself can result in serious physical and psychological harm and is most often imposed where consent is impossible to give. It is an issue of bioethics and of human rights, and it should be opposed on these grounds. It is estimated that two million women are affected by the practice annually. Often referred to as female genital mutilation, excision is mainly encountered in

---

* President of Rainbo, New York (United States).

1. Excision is the commonly accepted term for procedures that are technically known as 'female circumcision'. The term 'female genital mutilation (FGM)' includes excision and any other 'procedures involving partial or total removal of the external female genitalia or other injury to the female genital organ whether for cultural, religious or other non-therapeutic reasons' (World Health Organization, *WHO Information Fact Sheets*, No. 241, June 2000; see http://www.who.int/inf-fs/en/fact241.html).

twenty-eight African countries, in regions to the north and south of the equator and along the Nile valley, excluding the extremes of northern and southern Africa. It is also present among minorities in Asia, and has affected women who now live, either as immigrants or as refugees, in Europe, Australia, New Zealand and North America (Toubia, 1995). The ritual is practised by a minority of followers of many religions, including Christians of all denominations, Sunni and Shiite Muslims, Ethiopian Jews (the Falasha) and adherents of indigenous African religions.

The World Health Organization (1996) classifies excision into three general categories:

- type I: excision of the prepuce with or without partial or total excision of the clitoris;
- type II: excision of the prepuce and clitoris with partial or total excision of the labia minora;
- type III: excision of part or all of the external genitalia and stitching/narrowing of the vaginal opening (infibulation).

Objection to excision has often been based on the common physical complications that can result from the procedure, such as bleeding, infection, cysts, vaginal obstruction, etc. However, such arguments are problematic for two reasons. First of all, they infer that, in cases where no physical complications occur, the ritual is acceptable. Secondly, they implicitly suggest that the procedure should be performed by trained physicians or nurses under hygienic conditions. Indeed, such suggestions were made in Holland and the United States, but without success. Similar suggestions were actually followed up for a short period in Egypt in 1994, with serious consequences.

The objection to excision must rather be based on a bioethical argument. As indicated in the World Health Organization classification above, common types of excision involve the removal of healthy, functional organs. The clitoris and labia minora are the specialized sensitive sex organs of the female and their removal can interfere with a child's sexual development and later with her sexual enjoyment as an adult woman. The ethics of the health profession is based on the sworn oath to do no harm. But excision, in all its forms, clearly involves harm to the woman.

Figure 1. Areas of the world in which female genital mutilation (FGM) has been reported to occur

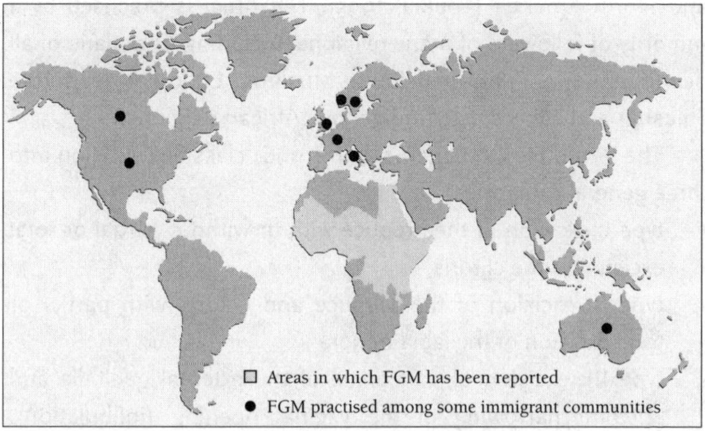

IMMEDIATE COMPLICATIONS

- Haemorrhage
- Shock
- Infection
- Urine retention
- Injury to adjacent tissue

LONG TERM COMPLICATIONS

- Bleeding
- Difficult micturition
- Recurrent urinary tract infections
- Incontinence
- Chronic pelvic infections
- Infertility
- Vulval abcesses
- Keloid formation
- Dermoid cysts
- Neurinoma
- Calculus formation
- Fistulae
- Sexual dysfunction
- Difficulties in menstruation
- Problems in pregnacy and childbirth
- Risk of HIV transmission
- Psychological and social consequences

Source: WHO, 'Women's health', WHO position paper for the Fourth World Conference on Women, 1995, p. 14.

From a human-rights perspective there are other points to consider. In its traditional setting, the ritual is most commonly performed on a child, who obviously cannot give consent. Excision is thus an infringement on the right of a female child to preserve her body integrity. Parental rights and wishes should not be considered in cases where physical damage is irreversible.

Bioethical and human-rights issues such as these were considered at several recent international conferences that condemned excision and called for efforts to stop the practice. These include the World Conference on Human Rights (1993), the International Conference on Population and Development (1994), the World Summit for Social Development (1995) and the Fourth World Conference on Women, Beijing (1995). The combination of documents published following these conferences constitutes a strong consensus of the world community that excision is a risk to reproductive and sexual health and a violation of human rights.[2]

Many countries and professional organizations have already taken a stand against excision. The United Kingdom, Sweden, Norway, Australia, New Zealand and the United States have all passed legislation against the practice. In Africa, Ghana passed a law against excision in 1994 and Burkina Faso did so in 1996; in Egypt, a Ministerial decree prescribing criminal penalties came into effect in 1996. Some international medical and health organizations have also taken policy stands against the practice, among them the World Health Organization, the International Federation of Gynaecology and Obstetrics (FIGO), the International Medical Women's Association and the International Nurses Association. In Europe and North America many national professional organizations have taken similar policy positions. However, none of these professional bodies has contemplated passing any professional regulation clearly prohibiting their members from performing the practice and prescribing penalties for

2. For example, see United Nations, 1993*a*, 1994*b*, 1995*a* and 1995*b*.

those who do not comply. Such regulatory measures from within the health and medical community would be binding for members in countries where excision is prevalent, and send a strong message to health professionals that their participation in this ritual would have grave consequences. Such actions are urgently needed since there is increasing evidence that physicians and nurses are conducting the practice,[3] often with financial gain as their motivation.

3. This trend was most prominent in Egypt, where a recent Demographic and Health Survey showed that while in the mothers' generation only 13 per cent of girls underwent excision by physicians, the number rose to 46 per cent in the daughters' generation (Moore, 1997).

Reproductive health

Attiya Inayatullah*

The attainment of reproductive health is hampered by a wide range of factors that are cultural, social and medical. Discrimination against women, which, as the previous chapter indicates, often begins in childhood, results in standards of nutrition and health care for girls that fall far below those of boys. When young women's general health is poor, the demands of pregnancy and childbirth have severe effects. For example, more than half of pregnant women in developing countries are anaemic, and anaemia is now believed to be a contributory factor in many maternal deaths (WHO, 1995*d*). Undernourished women often fail to gain sufficient weight during pregnancy, and each pregnancy may make the situation worse, especially when short intervals between pregnancies leave inadequate time to build up reserves again.

Early marriage, sometimes forced on young women, and early child-bearing are still common practices. Indeed, they are encouraged where a woman's status depends on her ability to

* President, International Planned Parenthood Federation, London (UK).

bear children, particularly sons. Women in these situations are subject to the well-known general risks associated with high fertility, risks that result from pregnancies occurring too early and too late in life, from too many pregnancies and from pregnancies that occur too close together. The situation is made worse when the decision to use contraception is not that of the woman, but of her husband or his family. Unsafe abortion, the last resort of women who have become unwillingly pregnant, carries high risks of infection, sterility and even death.

Lack of information and lack of access to proper health and family-planning services are two other factors affecting reproductive health. Some 350 million women are still without access to a full range of modern methods of family planning (United Nations, 1995*c*). According to one estimate, 120 million more women would currently be using a modern method of contraception if more accurate information and affordable services were easily available, and if their families were more supportive (ibid.). According to other estimates, 150 million women would like to delay or prevent a further birth but are not using contraception (Population Reference Bureau, 1997). In a number of countries, the consent of a spouse is needed before surgical contraception can be performed.

Figure 1. Unmet family planning needs: married women who want no more children but do not use any contraceptive method

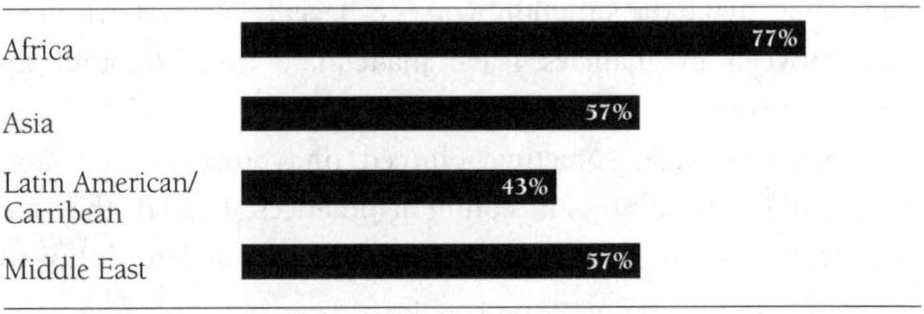

Source: WHO, 'Women's health', WHO position paper for the Fourth World Conference on Women, 1995, p. 14.

Further threats to women's reproductive health come from high-risk sexual behaviour and violence, including rape. When women are unable to negotiate the use of condoms with their partners, and many are in that position, they risk becoming infected with a sexually transmitted disease (STD), including the HIV virus. STDs disproportionately affect women: women seem to be more vulnerable and, because many STDs occur without symptoms in women, they are less likely to know they have an infection and seek treatment. The proportion of women affected by HIV/AIDS has been steadily growing, and has now reached 42 per cent of the total number of cases. The spread of HIV is particularly rapid among young women, who are also becoming infected at an earlier age than men.

National surveys from the Netherlands, New Zealand, Norway, the United States and Canada show that a quarter to a third of women have suffered childhood sexual abuse, with long-term health and emotional consequences (Population Reference Bureau, 1996b). The proportion of adult women who have been sexually assaulted is also high, ranging from 6 per cent in Moscow, to 15 per cent in Buenos Aires and reaching 22 per cent in Kampala. The implications for reproductive health include chronic pelvic pain, pregnancy complications, miscarriage, maternal mortality and, perhaps more significantly, adverse effects on psychological health. Since reproductive health involves the full physical and psychological well-being of the individual, such psychological instability must not be overlooked. Another act that often causes significant harm and has serious effects on reproductive health, as the case studies of the previous chapter attest to, is female circumcision.

It is clear from this small selection of statistics that all too many women are still not able to exercise their right to make decisions about child-bearing, let alone their right to sexual and reproductive health. In many parts of the developing world reproductive health care remains an unattainable luxury. Women who suffer are sometimes denied their rights not only by husbands or families, but also by governments that deny them the information and services they need to make free and informed choices to ensure their reproductive health.

Principles of reproductive health

While there is no official international code of ethics concerning reproductive-health issues, some fundamental principles have been recognized. In the preparation for the International Conference on Population and Development (Cairo, 1994), an alliance of women's groups (Women's Voices, 1994) prepared a list of reproductive-health principles which included the following: women must be subjects, not objects, of any development policy; population policies must be based on the principle of respect for the sexual and bodily integrity of girls and women; women have a right to information and services; men also have a personal and social responsibility for their own sexual behaviour; sexual and social relationships between women and men must be governed by principles of equality, non-coercion, mutual respect and responsibility; fundamental sexual and reproductive rights of women cannot be subordinated, against a woman's will, to the interest of partners and family members.

Research and its implementation

The spirit underlying these principles is crucial to scientific research, especially in the areas of contraception and infertility treatment. At present, most contraceptive methods are used by women. An important addition to the range of methods available is that of contraceptive implants, which are long-term and easily reversible. Nevertheless, there have been reports of women trying to remove the implants themselves: some had not been told that they had the right to request removal at any time. In other cases those who had been trained to insert the implants had not been trained to remove them.

There is only one widely available method of contraception that protects against both pregnancy and sexually transmitted disease: the condom; it must be used by men. For many years there has been a call for more research into contraceptive methods for men and, though some progress is now evident, as yet no safe, effective, reversible and acceptable method for men is available. This is an urgent matter, especially if men are to be encouraged to share the decision-making and the responsibility for the regulation of fertility.

In the comparatively new area of infertility treatment, a number of ethical dilemmas have arisen and more are on the horizon. First, people have questioned the legitimacy of medically assisted conception when children are available for adoption. Since techniques like *in vitro* fertilization (IVF) result in more multiple births, there is a growing need for an examination of the effects of these techniques. A study being undertaken in France on the psychological trauma experienced in later life by children born as a result of IVF was stopped before completion. Doubts have been expressed about the wisdom of making such a new technology more widely available before more research has been carried out. The long-term effects of infertility treatment on the health of the women also needs investigation.

Policy, services and the quality of care

Ensure that all health services and workers conform to human rights and to ethical, professional and gender-sensitive standards in the delivery of women's health services aimed at ensuring voluntary and informed consent; encourage the development, implementation and dissemination of codes of ethics guided by existing international codes of medical ethics as well as ethical principles that govern other health professionals
(Fourth World Conference on Women, 1995).

The spirit of the principles mentioned in the passage cited above must be respected in policy-making and when providing health services. Indeed, it is in these areas that abuses can most easily occur. Issues such as enforced sterilization have been widely condemned, but there are others requiring examination: for example, the use of targets or incentives too generous to be turned down by the impoverished. An over-enthusiastic family-planning service provision may border on coercion, and while every woman should have the right to control her own fertility, enforced fertility control can become an oppressive, rather than a liberating, experience.

The quality of care available throughout the delivery of health services is crucial to the protection of human rights. The quality of care must start with the user, with free and informed choice as an underlying objective. The view that some areas of reproductive health have been made too inaccessible has been voiced for some time. One example: certain methods of contraception are only available through clinical services or by prescription. Another is the over-emphatic advice given to women whose families are 'complete' to have hysterectomies. In the 1980s a third of women in the United States and a quarter of women in Australia aged between 50 and 54 years underwent hysterectomies, compared with 12 per cent of women in the same age-group in Sweden (Santow and Bracher, 1992). The prevalence of deliveries involving caesarean sections in some developed countries may reflect the convenience of health providers rather than the wishes of women. All such interventions move birth into a clinical setting and enhance the role of physicians, policy-makers and eventually lawyers, while at the same time diminishing the woman's control of her own body. Many of the most recent advances in contraception necessitate skilled providers, which makes it even more important that acceptability as well as effectiveness be evaluated.

Elements of reproductive health

Fertility and family planning

Family planning can be the first step towards empowerment for many adult women. It may be the first time a woman makes a choice of her own, and it may be her first experience of being treated with dignity and respect. It can start the process of transforming power relations in families and communities. With improved health and fewer children, women may choose to take advantage of other opportunities – to continue education, earn an income or participate more fully in their communities. Furthermore, access to family planning indicates an egalitarian society where the rights of women are valued and where there is a greater likelihood that their reproductive health will be cared for.

The essentials for family planning are that it must be accessible, safe, effective, cheap and culturally acceptable. The judgement of suitability must be made by the user of the service and not the provider. The validity of government population policies rests upon their capacity to improve the quality of life of the nation's citizens, and the understanding and agreement of those citizens.

Some 510 million couples currently use contraception throughout the world. In many regions the choice of methods is limited, not only because of economic constraints, but also because of social and political reasons. For example, oral contraceptives may not be offered where providers consider it difficult for women to adhere to a strict pill-taking regime. During the 1970s the Indian government emphasized sterilization for couples who had two or more children; other, less definitive methods could have been proposed but were not.

Since contraceptive failures cannot be eliminated, the ethical imperative of providing and promoting emergency contraception should be stressed. Despite fears that the availability of emergency contraception would reduce the use of other contraceptive methods, generally this claim has not been substantiated.

The health risks of contraception must also be taken into account when reviewing the reproductive-health benefits of fertility management. Reproductive health may be put at risk by the very techniques that are designed to improve it. For example, a known side-effect of the intra-uterine device is to increase menstrual flow; but this may be unacceptable where women are already anaemic or where the discharge of blood has a particular cultural significance. However, the health risks of pregnancy are often far greater than any risk from contraception, and this has to be weighed in each individual case.

There also exists the potential for abuse of contraceptive technology. Methods such as implants, which act over a long period of time, are especially open to manipulation by the authorities: for example, their presence may be considered necessary before a woman can receive some welfare benefits. Women might be given fertility vaccines without their knowledge or consent, thus violating their basic human rights. More difficult ethical questions are raised where fertility control is imposed on some women because of judgements about their mental status. For example, is it right for handicapped women to be given contraceptives when they are not able to make their own informed choices?

Infertility and childlessness

Involuntary childlessness signals poor reproductive health. It may be due to difficulties in conceiving, carrying a pregnancy to term or high infant-mortality rates. Eight to 10 per cent of couples in the world – an estimated 50 to 80 million couples – are infertile (WHO, 1992). Pregnancy loss can be caused by endemic infection, syphilis, malaria, malnutrition or obstetric problems. In Africa, where up to 30 per cent of couples may be infertile in some areas, nearly two-thirds of infertility in women is the secondary result of infection by STDs, pelvic inflammatory disease (PID) or infections associated with pregnancy complications (Sciarra, 1994). Changing sexual habits and effective early treatment of disease could reduce infertility generally, and better medical care could reduce the infertility that results from postpartum and abortion-induced infections.

Infertility is often experienced as a personal tragedy, especially since it is usually the woman who is blamed for failing to conceive, even though much infertility is due to the male partner. In many societies a husband may leave a 'barren' wife for another woman if she fails to conceive or produces only female babies. The treatment of infertility can be controversial. Some forms can be dealt with fairly simply, with the administration of antibiotics to clear up infections or of hormones to deal with the failure to ovulate. Sperm donation is an option for male infertility. Two technically sophisticated methods of treatment are *in vitro* fertilization (IVF) and gamete intra-Fallopian transfer (GIFT), although there are risks associated with these techniques, notably the risk of multiple pregnancies (see previous chapter, 'Infants, children and adolescents', p. 40). Other problems likely to arise are pre-eclampsia, prematurity and neonatal mortality. A world survey in 1987 showed that 24 per cent pregnancies after IVF or GIFT were multiple ones. A woman may have IVF treatment with eggs donated by another woman, or there may be surrogacy, where a woman gestates the embryo of another couple. And there is the further issue raised by the fact that these technologies are inaccessible to many because of financial reasons. For example, a pregnancy following treatment of infertility costs four to five times as much as a 'normal' pregnancy. Some sort of rationing may therefore be needed, and at present this is usually based on age – older women generally have poorer IVF and GIFT outcomes at every stage – or on marital status. The roles, and claims, of the state, of health and other professionals and of the individual in making these decisions are still hotly debated.

Safe motherhood and maternal morbidity and mortality

Pregnancy and childbirth represent important periods in most women's lives. Women also normally play the more significant role in child-rearing. The right to safe motherhood is far from being a reality, and throughout the world many women give birth in unsatisfactory conditions.

For each of the 500,000 women who die every year because of complications experienced during pregnancy, UNICEF (1996) estimates that a further 30 million suffer from lifelong disabilities such as sepsis, reproductive tract infections and infertility. It is calculated that infant mortality could be reduced by as much as 30 per cent in some countries, including Pakistan and Egypt, by using family planning to gain optimum conditions for child-bearing. Access to emergency obstetric care is the single most important reason why maternal mortality has almost disappeared in the developed world. Maternal morbidity could be reduced by providing well-trained midwives and efficient care during delivery.

A number of social factors also impact on maternal and infant mortality and morbidity. In Australia, the number of maternal deaths among non-indigenous women declined by almost two-thirds from 1970 to 1990. However, among Aboriginal and Torres Strait Islander women there has been relatively little change in the number of maternal deaths during this period. About 2 per cent of all births in Australia are to Aboriginal and Torres Strait Islander women. But these women accounted for 30 per cent of all deaths attributed directly to obstetric complications from 1988 to 1990 (Office of the Status of Women, 1994).

Figure 2. Maternal deaths per 100,000 live births by region, 1988

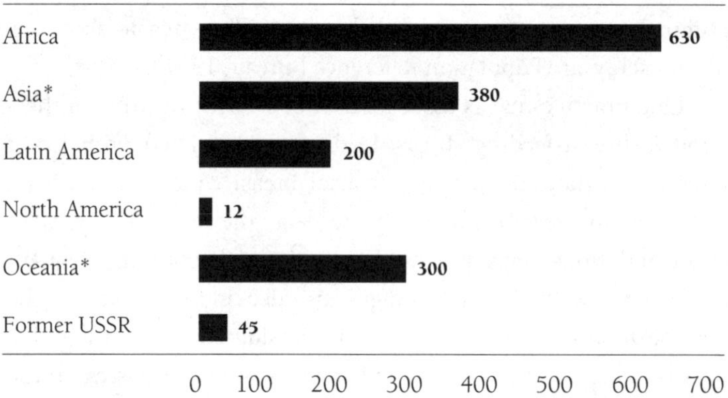

* Excluding Australia, Japan, New Zealand

Source: WHO/MSM database, early 1990s; reproduced in WHO, Women's health, WHO position paper for the Fourth World Conference on Women, 1995, p. 7.

Abortion-related deaths account for roughly one third of maternal deaths each year. Unwanted pregnancy is a specific reproductive-health risk for women and their families, and it remains one of the main deterrents to achieving a satisfactory level of reproductive health. Abortion remains a highly contentious issue, and debates often centre around the exact moment of conception, or differing ideas about the relative rights of mother and embryo. In 1994, 173 of 190 countries worldwide permitted abortion to save a woman's life; 81 allowed abortion in cases of rape or incest; and 55 allowed it for economic or social reasons (United Nations, 1994*c*).

The law, however, is often an irrelevant issue for a woman seeking an abortion. Moreover, the legality of the practice provides only a small indication of its safety. Even where abortion is legal, safe abortions may be inaccessible because of cost. As a result, women, especially the young and the poor, seek untrained practitioners. Similarly, where abortion is illegal, those who can afford to pay a large sum of money may be able to obtain a safe abortion from a competent provider. But women's reproductive health can never be fully assured if women or their doctors are forced into criminal actions. It is estimated that 20 million unsafe abortions occur each year (Population Reference Bureau, 1997).

One practice that is associated directly with healthy motherhood is breast-feeding. Breast-feeding is associated with lower rates of ovarian and pre-menopausal breast cancer in mothers, and frequent breast-feeding helps to delay the return of menstruation and thus helps the mother to avoid becoming pregnant again too quickly. Its advantages are also indisputable for the new-born baby. Breast milk alone is considered the best possible nutrition for an infant in the first months of life. It passes on the mother's protection against certain illnesses and it helps to prevent diarrhoea, which can easily occur among bottle-fed babies in communities without clean drinking water. However, less than half of the babies worldwide are exclusively breast-fed in the first three months of life (UNICEF, 1997). Breast-feeding has declined for a variety of reasons, including the rise of new lifestyles that have reduced the traditional importance given to it and the increased proportion of women in the formal workforce.

STDs, including HIV and reproductive tract infections

There are an estimated 333 million new cases of STDs each year (WHO, 1995*a*). There were 3.1 million new infections with HIV in 1996 (UNAIDS, 1996). The presence of one STD increases rates of infection by other STDs. They cause pregnancy-related complications, sepsis, spontaneous abortion, premature birth, stillbirth and congenital infections. STDs such as HIV may be passed from mother to child during pregnancy and lactation. Almost two-thirds of cases of infertility are attributable to STDs, and 17 to 40 per cent of all gynaecological admissions are due to pelvic inflammatory disease (PID) (United Nations Population Fund, 1997).

Overall, the burden of STDs is five times greater for women than men. This is partly because of the greater mucosal surface exposed to a greater quantity of pathogens during sexual intercourse, and because women are often more vulnerable and generally weaker as a result of pregnancy, childbirth and lower nutritional levels. Difficulties in diagnosis are especially acute in women, as up to half of STDs in women are asymptomatic. Furthermore, up to 30 per cent of HIV infections occur in women whose only risk behaviour is sexual intercourse with a single male partner, who may be their own husband, who had unprotected sex with other partners (WHO, 1995*c*).

Services available are often extremely limited since STD management is rarely provided as part of an integrated approach to women's health needs. Screening for most STDs is comparatively expensive and requires sophisticated equipment. While advances in HIV/AIDS therapies make headlines, for most of the world the cost of such treatment remains prohibitive. Some progress has been made in developing countries with treatment that prevents transmission from mother to child. However, given the slow pace of the development of a vaccine, prevention is still the only realistic strategy.

Case study

Hope for South African mothers and newborn babies

Sister Christa Mary-Jones*

The development of health-care services and the beginnings of midwifery

The Decentralised Education Programme in Advanced Midwifery (DEPAM) began officially in 1989 at McCord Hospital in Durban, South Africa. Its conception dates back many years to the remote and under-serviced rural areas of KwaZulu/Natal.

Formal health-care services were first brought to the indigenous peoples of South Africa by missionaries. A hundred years ago the inhabitants of the region faced poverty and a total lack of facilities to assist the ill or promote healthy living. Missionaries encouraged local people to participate in health-care services, and this encouragement proved successful. The first African nurses were trained in mission hospitals, and they passed their examinations in city hospitals, in a language that was not their own and without the benefits of libraries and sophisticated teaching aids. In the ensuing years of racial division, mission hospitals were taken over by the government and many of them subsequently deteriorated seriously.

Doctors were reluctant to work in areas that lacked sufficient infrastructure; these areas suffered from communication difficulties and were associated with low wages and low social status. However, a small team of dedicated rural missionary doctors continued to provide service in difficult situations. An emphasis

* St Mary's Catholic Mission Hospital Trust (South Africa).

on teamwork, with participation from local nurses and midwives as well as from the community at large, was an integral part of their method. As a result, the doctors came to appreciate, and subsequently promote, the skills of African nurses and midwives.

One such doctor was J. Larsen. Dr Larsen began upgrading the skills of midwives in the Nqutu region of KwaZulu by offering training over a period of six months that was designed to develop skills in the following areas: the screening of pregnant women to detect general problems; the administration of early treatment or immediate referral to a hospital; expertise in managing obstetrical emergencies; a correct and efficient use of forceps, vacuum extractors and symphysiotomies; care of the newborn and management of emergency conditions; and awareness of methods of family planning and holistic health care. Dr Larsen referred to his midwives as 'super midwives', and they provided service in the Nqutu region and beyond.

The idea of developing the skills of midwives was eventually accepted more generally. The South African Nursing Council officially recognized the Diploma in Advanced Midwifery and Neonatal Nursing Science. This was listed as a postgraduate training programme available to professional nurses who had already undergone five years of nursing/midwifery education. 'Advanced Midwifery' took a year to complete and the midwife had to leave her home and her workplace to study in the major urban academic hospitals in which it was offered.

These institutions had the benefits of having the best technology and skilled personnel available, but the training given to the midwives in this setting tended to be based on the medical model and concentrated on curative aspects rather than community-based care. As a result, the Advanced Midwife graduates became proficient in managing problems in ideal settings with optimum human and technical support. However, they were not adequately prepared to work in the remote and often neglected clinics and hospitals to which they returned after this year of training. Many struggled to function and found it difficult to implement change and, in frustration, some left the rural areas and returned to take up nursing in the city. Consequently, the maternal and neonatal health care of the rural people remained poor.

The Decentralised Education Programme in Advanced Midwifery (DEPAM)

Recognizing that rural midwives were often unable to leave home to study for a year in an urban hospital because of family commitments, that many of those who did failed to return to rural areas to practice, and that theoretical acumen was not necessarily of primary importance for midwives working in remote, less-than-ideal circumstances, the idea of the DEPAM was developed in the late 1980s by midwives and doctors involved in maternal and neonatal care in the under-served rural areas of South Africa. The first DEPAM was held at McCord hospital under the direction of Dr P. Garde and Ms N. Mzolo, and supported by Professors S. Ross, M. Adhikari and W. Loening from the University of Natal, among many others.

DEPAM was set up as a community-based adult-education programme that sought to train midwives working in difficult areas to provide the best possible care and to evaluate objectively, on an ongoing basis, the services they were giving. Emphasis was placed not only on technical skills and competency, but also on initiative, responsibility, assertiveness, dedication, self-directed learning and the ability to solve problems and adapt to change. Teaching skills associated with quality assurance methods, conducting meetings and evaluating maternal and perinatal mortality and morbidity was also included in the course.

In order to emphasize the need for communication with the community at large, each advanced midwife must undertake fifty home visits in the area where she works and must formulate a community diagnosis. She is trained to listen to the members of her community with the aim of determining what they perceive as their problems and what they feel are the solutions. During these visits the midwife carefully observes general conditions of maternal and child health and begins to establish the type of personal contact that is crucial to a community-based programme.

Selection of candidates

The selection of candidates is not based solely on academic merit, nor on the recommendations made by nursing authorities. Rather, it involves considerations in keeping with the general aims of the programme as outlined above. Candidates must demonstrate not only overall competence, but also an ability to lead, provide innovative solutions and study independently. Assessments of the candidate made by her peers and by the patients she cares for are examined. Candidates are then interviewed in order to establish their commitment to a future career as a clinical nurse specialist, and they must agree to remain in the rural areas after their training and to participate in the development of the whole health-care team.

The programme of education

This course is organized on a part-time basis over a period of 18 to 24 months. This includes six weeks of annual group study, practical and theoretical evaluations, the organization of regional perinatal-mortality meetings and in-service education sessions. Should the student need experience in intensive care units or specialized maternal- or neonatal-care units, this is organized with the referral hospitals in her area. Social workers, community leaders, formal and informal health-care providers, traditional birth attendants, herbalists, healers, paramedical staff, doctors, nurses and established midwives all participate so that students can develop a wide network of relationships. Great importance is placed on developing the ability to lead and to communicate with the community and with the health-care team.

Problem-based learning is considered as the cornerstone of the programme, and each encounter with women and newborn babies provides stimulus for continued education. Clinical supervisors from rural hospitals or clinics participate in all areas, and trained advanced-midwife facilitators co-ordinate the course. Self-assessment as well as facilitator counselling and evaluation continue throughout the course and culminate in an approved examination process that includes the presentation of research, written examinations and practical assessment.

Facilitator training

One of the main goals of DEPAM is to make the programme available to midwives throughout the country. By training skilled, advanced midwives as facilitators and by encouraging them to start independent DEPAM training centres in other provinces of South Africa, it is hoped that the benefits will spread nationally. McCord Hospital continues to train candidates as facilitators, and at present the programme is under way in six provinces.

Problems

Although the development and expansion of the programme continues apace, a number of problems and drawbacks have been encountered. For example, students still experience stress from leaving their families even for short periods of time, and this remains a general problem. Studying at home while working full-time is also found to be difficult, as is the return to studying by women who have not applied themselves in this capacity for several years. Nurse managers, burdened by a shortage of staff, were initially glad to hear of a programme that did not require staff leaving the hospital or clinic for a year; however, they have often been reluctant to free students for periods of study or for additional experience or community work since they still have them on their staff complement, and this has caused tension between the students and the nursing authorities.

In addition, DEPAM students themselves thought that they understood the concepts of self-directed learning and adult education before they started the course, but have struggled with group dynamics and facilitation that favours men. Finally, authorities, both legislative and educational, appear to be wary of innovative change, and the dissemination of the programme throughout the country has been hindered by bureaucracy. Further financial support will be necessary to set up resource centres, provide transportation to remote areas and pay the salaries of facilitators.

Has DEPAM made a difference?

The training of DEPAM midwives is based in the hospitals and clinics in the regions where the women live, alongside the resident doctors and nurses. This has two advantages for prospective midwives: first, they gain respect while still training; and secondly, they become familiar with their working environment early on. Consequently, DEPAM midwives have generally found it easy to bring about change in the hospitals and clinics where they work.

Members of the community have expressed general satisfaction with the work of DEPAM midwives. The distance between formal and informal health-care providers has diminished, and DEPAM midwives have become involved in traditional birth attendant training and community health worker training in their areas. These women have also participated in projects for the general good of the community. For example, they have aided in building roads and bridges in the Mpola and Ixopo regions, where women and children were dying in childbirth because of their isolation, and they have learned how to build concrete water tanks and pit latrines. They have also entered into dialogue with warring tribes to allow for the care of women and children.

DEPAM midwives have influenced policy and protocol by becoming involved in regional and local health-care planning. When clinics and maternity wards are being designed and built, they have become involved in planning them so that women will feel respected, accepted and comfortable and thus be encouraged to use them. Prior to 1994, they challenged the health-care service of South Africa, with its emphasis on curative health, and they have been enthusiastic about the primary health-care outreach services of the 'new' South Africa.

Special attention has been given to preventing maternal debilitation and mortality. Malaria, HIV/AIDS, sexually transmitted diseases, anaemia, parasitic infestations, adolescent or grandemultiparous pregnancies, pregnancy-induced hypertension, eclampsia, pregnancy-related haemorrhage, and the problems of women and children living in slums and squatter settlements have all been targeted.

These are some of the advantages that have been observed so far. More are hoped for. The programme is flexible, requires a minimum investment in new technology and is suitable for developing countries and regions. Its spread to other parts of South Africa, and even beyond, would be welcome.

Case study

Deferred parenthood in the Netherlands

R. M. den Hartog-van Ter Tholen*

In December 1996, the Dutch Equal Opportunities Council published a report on deferred parenthood. As part of its preparations for this report, the Council had commissioned research on the social consequences of couples deferring the birth of their first child.

* Curative Somatic Care Department, Medical Ethics Division, Ministry of Health, Welfare and Sport (The Netherlands).

This article is a summary of the Council's report[1] and the preparatory research report,[2] supplemented by some statistical information from an article by Van Balen, Verdurmen and Ketting, published in *Human Reproduction* in 1997 under the title 'Age, the desire to have a child and cumulative pregnancy rate'.

Postponement

Since the advent of reliable methods of contraception, parenthood has become a matter of choice and hence a question of planning and decision-making. A growing number of couples are deferring the decision for ever longer periods and therefore postponing the birth of the first child. As a result, the mean age at which women give birth to their first child is steadily rising. In the Netherlands, the mean age of mothers giving birth for the first time was 24.3 years in 1970, 27.6 years in 1990 and 28.4 years in 1994.[3] In 1970, most women (63 per cent) were under 25 when their first child was born, 27 per cent were aged between 25 and 29 years, and only around 10 per cent were over 30. In 1990, however, only one-third were under 25, while a quarter had reached the age of 30 when their first child was born. The mean age of mothers at the birth of the first child is higher in the Netherlands than anywhere else in Europe (Table 1). As the table shows, there is a similar trend in other European countries.

1. *Het late ouderschap: over uitstel en afstel* (Late parenthood: postponement and renunciation).
2. *Uitgesteld ouderschap* (Deferred parenthood).
3. According to information provided privately by GCN Beets of the Netherlands Interdisciplinary Demographic Institute, the figure for 1997 is 29.0 years.

Table 1. Mean maternal age at birth of first child in the member states of the European Union

	1970	1980	1990	1994
Austria	23.7	24.3	26.1	25.9
Belgium	24.3	24.5	26.4	
Denmark	23.7	24.6	26.4	27.2 [a]
Germany			26.9 [c]	27.5 [a]
Former BRD	24.3	25.2	26.9	27.6 [a]
Former DDR	22.5	23.3	24.7	26.2 [a]
Finland	23.7	25.7	26.8	27.4
France	23.8	24.9	27.0	27.4 [b]
Greece	24.0	23.3	24.7	26.1
Ireland	25.3	24.9	26.3	26.6 [a]
Italy	25.1	25.1	26.9	27.4 [a]
Luxembourg				
Netherlands	24.3	25.6	27.6	28.4
Portugal	24.4	23.6	24.7	25.4
Spain		24.6	26.5	27.2 [b]
Sweden		25.5	26.3	27.0 [a]
United Kingdom	23.5	24.4	25.5	26.2 [a]

a. 1993, b. 1992, c. 1991

Source: Council of Europe, 1995.

The motives

There are various reasons for deciding to defer having a child. These include the parents' desire to complete their education, the loss of human capital (the total productive knowledge and skills of the individual), the reduction in lifetime income if a parent leaves paid employment or devotes fewer hours to it, and the desire to achieve a certain level of affluence before starting a family. In the Netherlands, however, the main reason for postponement seems to be the difficulty of combining child care with paid employment.

Women's participation in the labour market

Until very recently, women's participation in the labour market was very low in the Netherlands. In 1960, only 26 per cent of women were in paid employment. By 1994, the figure had increased to 54.4 per cent. Many factors are influencing this trend. For instance, the level of education, since more highly educated women tend to be more active in the labour market. Many women stop work when they have children, or switch to working part-time; so deferring the birth of a child allows more women to work longer in their earlier years. In addition, experience has shown that increasing the number of places in official child-care programmes increases the number of working women. However, a considerable proportion of the Dutch population still finds child care outside the home unacceptable. For this reason, it is not in itself sufficient to increase the number of child-care places. In spite of the trend for women to work, their relatively low pay nevertheless makes it less attractive for them to seek or remain in employment.

According to the study by Van Balen et al., the age at which couples first actively desire to have children is a function of their level of family income, professional status and level of education. The study reveals no significant correlation with religion, region, level of urbanization or number of incomes per family.

Medical consequences of postponement

As women grow older, their fertility declines and the risk of medical complications during pregnancy increases. Each month up to the age of 30, a woman of normal fertility has around a 20 per cent chance of conceiving. By the age of 35 the rate has declined to 10 per cent and by the age of 38 to only 5 per cent. This also means, however, that a woman of 35 still has a 75 per cent chance of conceiving within 12 months and an 80 per cent chance of doing so within 24 months. Van Balen et al. concludes that there is a turning point at the age of 33, after which there is a sudden sharp diminution in the chance of achieving a first pregnancy that will go to term.

For this reason, the trend towards deferred parenthood is producing an increase in the number of patients seeking fertility treatment and, despite these treatments, a growing number of permanently childless couples. Artificial reproduction treatments are associated with considerable psychological pressures, and the hormone treatment which must usually accompany it is not without risk. In addition, the multiplicity of medical techniques available means that the hope of conceiving is now sustained for much longer and the process of grieving over childlessness is postponed until much later. Many couples experience their inability to have children as one of the worst things that has happened to them.

Higher maternal age also increases the risk of complications during pregnancy. The most frequent of these are toxaemia, gestational diabetes and high blood pressure. The rates of miscarriage and Caesarian section also rise, as does the risk of giving birth to a child with a congenital abnormality. This increases the demand for prenatal diagnostic testing, which is in turn associated with psychological effects. One third of all women who decide to terminate a pregnancy following an adverse diagnosis subsequently experience prolonged depression.

The rate of stillbirths and infant mortality is higher among mothers aged 35 and over than among those aged between 25 and 35. Infant mortality immediately before or during delivery or in the first week of life increases from a mean figure of 8 per thousand births for mothers aged 25–29 to 9.6 per thousand for those aged 30–34 and 12.4 per thousand for mothers aged 35 or over. This is connected with the higher rates of premature delivery, multiple pregnancy and congenital abnormality for this older age group.

Financial consequences

Because of the increased demand for reproductive technology and consequently for medical and psychological support, the deferment of parenthood also has considerable financial consequences. For this reason, the Dutch Equal Opportunities Council feels that investment in measures and facilities to make it easier to combine work and family life might eventually pay for itself.

Facilities to make it easier to combine work and family life

There are various ways of making it easier to combine work and family life. First and foremost among these is to provide opportunities for part-time or flexible working. It would be helpful to create a statutory right to such working arrangements; however, a bill to this effect failed to attract sufficient support in the Dutch parliament. A second possibility is to create an entitlement to various types of leave: pregnancy and maternity leave, extra leave during the first year of the child's life and special leave in case of family illness. In the Netherlands, women receive 16 weeks of paid maternity leave around the time of delivery. Extra leave during the early years of the child's life is available to government employees on a partially paid basis, but special leave in case of family illness is almost unknown in the Netherlands.

In practice, these types of leave pertain only to women. There is a danger, therefore, that employers may become reluctant to take on women because of the greater risk of long-term absence. In that case, such arrangements would tend to reinforce rather than reduce the present inequality between men and women in the labour market.

A third possibility is to offer tax facilities and family allowances. In the Netherlands it is virtually impossible to deduct child-care costs from taxable income.[4] Parents receive child benefit for each dependent child. The amount paid is the same for all children in the age group concerned. Finally, child-care provision is important. In the Netherlands, the vast majority of child care (87 per cent) is provided informally within the nuclear or extended family or by neighbours. Formal paid child care accounts for only 13 per cent of the total. The organized child-care sector (crèches, official childminders and out-of-school facilities) provides places for only 2 per cent of children under the age of four and 3.7 per cent of those under the age of six. Child-care problems do not cease when children go to school. School hours do not correspond to the normal working day. Many primary schools (for children aged 4–12) now provide facilities for children to stay at school during the lunch break, but Wednesday afternoons (when the children are free) and holidays remain a problem. Places in out-of-school care are available for only a very small proportion (3 per cent) of children under the age of 12.

Conclusion

The Dutch Equal Opportunities Council recommends that the personal risks and social consequences of deferred parenthood should be publicized, and advocates improved facilities of the types described above to make it easier to combine work and family life.

4. OECD Employment Outlook 1995, OECD, Paris, pp. 171–200.

Towards men's participation in reproductive health

Roland Edgar Mhlanga*

When the International Conference on Population Development (Cairo, 1994) emphasized the need for women's empowerment, the intended implications were far-reaching. It was hoped, first of all, that women's empowerment would promote the improvement of reproductive health in both men and women and, secondly, that it would enable couples to decide freely whether, when and how often to reproduce.

Since women make up the majority of those who suffer reproductive health problems, most strategies so far have been directed only towards them, in spite of the fact that choice of method of method of contraception is often determined exclusively by the male partner. The male condom, the rhythm method, coitus interruptus, coitus interfemoris and abstinence all require the direct involvement of the man. Even 'modern' methods intended for use exclusively by the woman, including hormonal treatment and the new female condom, also require a certain degree of co-operation on the part of the male. Most

* Director, Maternal, Child and Women's Health, Department of Health, Republic of South Africa.

methods are thus subject to man's approval, including contraceptives that are injected and 'barrier' methods. Moreover, men also determine the kind of behaviour that promotes or subverts reproductive health. An example is the practice known as 'dry sex', during which there is greater danger of abrasion of the female genitalia. Some males prefer this kind of sex in spite of the fact that it increases the possibility of the transmission of infections, particularly HIV, through harm caused to the genitalia.

Considering these facts, it is imperative that men be made active partners in the promotion and preservation of women's general reproductive health. And with this goal in mind, we must first respond to a traditional misconception: it is quite simply not true that men have no interest in family planning or in women's reproductive health. For example, men in sub-Saharan Africa are responsible for decisions concerning the number of children a couple will have (Muhondwa, 1996; Terefe and Larson, 1993); and projects in both developing and developed countries show that if men are given the chance they will be supportive of women's reproductive health (Armstrong, 1986; Sachs 1994).

Ignorance is the true cause of the apparent lack of men's interest in preserving women's health. But men are interested in knowing how different methods of contraception work. This information is shared in places where men gather and, often, those that are considered more knowledgeable will be given a hearing. Groups of young men inevitably gather around the man who appears to be an expert. Given this pattern of behaviour, the question we must now ask is the following: which strategies can be used to capitalize on this interest and to ensure that men are given the information they need in order to take a greater responsibility for reproductive health?

Intervention

In order to bring such changes about, the following areas all need to be addressed: government decision-making, health services, education and research. Only when the commitment of government and other institutions is co-ordinated in a coherent and effective manner will people recognize their role in promoting responsible, healthy sexual behaviour.

Government decision-making

Most government policies are formulated and finalized by men, often with the almost total exclusion of women. The International Conference on Population and Development noted that men play a key role in bringing about gender equality. Therefore, since men occupy many influential positions within government, their professional commitment to promote the respect for women's human rights and the advancement of reproductive rights as well as gender equality can make a difference in changing the perception of their fellow citizens.

Health services

Although in the health services of many countries female nurses are in the majority and make up the major portion of health workers, they do not have the powers of doctors, who are mostly men. It is thus men who influence how issues concerning women are defined and addressed, and it could be argued that in order to improve the health care of women, health services need to be made gender-sensitive.

For many years, in both developed and developing countries, family-planning programmes have concentrated directly on women but very little on men. With respect to this service, the health-care system needs to take into account to a greater extent the whole family, including the adult men. Community-based initiatives are more likely to involve other members of a family, and experience shows that while these initiatives are more difficult, the prospects of sustainability are higher.

In treating women's reproductive health, the health services of many countries have consistently demonstrated insensitivity, and in many places they continue to do so. In investigating infertility, for instance, the woman is the prime suspect; only after all other things have been excluded is it considered appropriate to investigate the man. Examinations of women usually include invasive procedures and risks due to the use of anaesthetics. These habits prevail despite the fact that the male in most countries is responsible for up to 40 per cent of the causes of infertility, and that it is easier and less invasive to investigate the man. Health workers should be made aware of these facts and habits should be changed in order to take pressure off the woman.

During the counselling for genetic disorders that health services provide, it is often apparent that men hold women liable for such reasons as the following: (1) transmission of any infection or disease to the child; (2) assuming all the care of a child with a disability, whether it be congenital or an acquired deformity or a handicap; and (3) infidelity, apparently as a result of having a child with a disability, for, as the complaint often goes, the child cannot possibly be the man's since he assumes himself to be healthy and well. These are often cited as reasons for the breakdown of families when a child is born with a deformity or disability. This leads to added hardships on the woman, who has to fend for a family alone. Although women tend to move into paid employment after the collapse of a marriage, they are still likely to be left with a much lower disposable income. These considerations should be taken into account by health-service workers when genetic counselling is provided.

The health services also have a crucial role to play in the prevention of sexually transmitted diseases and in the promotion of healthy sexual behaviour. Many sexually transmitted diseases, for example, gonococcal urethritis and syphilis, produce visible symptoms in males more often than in females. The woman may be infected and not even know it. For women, however, the damage caused may be irreversible, often leading to infertility. In many societies such infections are looked upon as diseases of shame. Men seek treatment with or without the casual partner and very rarely with the stable partner, especially when the man suspects that he contracted the infection elsewhere; indeed, the onset of a disease usually spells the end of a relationship, with no commitment on the part of one partner to the other's health and well-being. In order to help couples address such problems effectively, health-care strategies must be developed to reduce stigmatization and to encourage men and women to seek treatment together.

Education

Education must be provided with the intention of protecting the woman, giving her the information she needs in order to look after her own interests, whatever they may be. Indeed, there are a great many issues that education must encompass. Two of the areas where education is crucial, which have been discussed above, are government and the health services. First of all, politicians need to be trained to play a more supportive role in promoting women's health, and thus gender-sensitive training should be provided at all levels of government. Secondly, in the health services there is a need for a retraining of workers concerning gender issues in order to avoid some of the problems discussed earlier in this chapter.

Moreover, the education of men and women generally is important. This is just one way to help men become more supportive of women and more understanding of the issues they face. The gender-sensitive education provided specifically for women is intended to help them communicate their desires and expectations in relationships and in society at large.

Education should also be provided on the many different types of gender-based violence and how to prevent them. Indeed, men are by far the perpetrators of violence, and various studies have shown that women and girls are frequent victims of acts committed by family members (Heise, 1994*a*; Heise et al., 1994). In a variety of different situations, which can range from a young woman of little financial means wishing to further her education to a prostitute who might unknowingly contract HIV, women are often coerced, with money or other rewards used as incentives. Other types of gender-based violence may even be culturally ingrained. For example, although the practice of attributing a dowry to a bride may at one time have been considered noble, in modern society it often has abhorrent consequences (see 'Infants, children and adolescents', p. 45). With respect to all of these situations, the education of boys and young men might go a long way to changing such patterns in male behaviour.

Research

There are many issues to be addressed concerning current research. Present research is limited, for example, in that it has not attempted to look at the reasons for men's lack of interest in reproductive health. In the future, useful research will recruit men as active participants. The general state of women's health also suffers because drug research is conducted mainly with white males of Caucasian descent. No allowance is made, for instance, for the variations in hormonal profile, body mass or circulating blood volume. It would be beneficial if in the future such research were more balanced.

Promoting understanding

Perhaps the most general issue, and perhaps indeed the most important one overall, is the promotion of understanding between men and women in the area of reproductive health. The terrain is vast. Men, for example, are often ignorant of the effects, side-effects and contra-indications of contraceptives. It is inconceivable to claim any form of advancement if such an issue is not addressed through education based on the principle that family-planning programmes have not been sufficiently adapted to suit the needs of the male client. By way of closing this chapter, I will suggest the following ways in which such understanding could be promoted more effectively:

- The use of condoms could be interpreted as indicating that men are prepared to participate in women's reproductive health; however, there is a need to develop accurate indicators of the efficacy of our information and communication materials in order to evaluate the latter properly.
- Methods of contraception that take into account variations in the age and cultural background of the user need to be developed.
- In certain traditional schools, where time-honoured rites and practices are taught, as well as in certain religious institutions, there is a need for well-designed materials to teach what types of contraception are acceptable within that system.
- Sex education should be designed to contribute to improving mutual understanding and respect between men and women.
- The consistent use of contraception, even those methods that are not very reliable, will lead to better results than the sporadic use of reliable methods. For example, the rhythm method will have a more positive result than oral contraceptives that are not taken regularly. In general, the best approach is to support the methods that people are comfortable with.
- Calling on men's sense of responsibility is the only solution to the feeling of helplessness that many women feel at protecting themselves against infection from sexually transmitted diseases. It is fashionable for HIV educators to maintain that condom use does not reduce sexual pleasure. However, many men and women admit, on the contrary, that condoms diminish sensation during sexual intercourse. It would be more useful if health workers also admitted this while at the

same time underlining the benefits of the condom. The aim is to make people more comfortable with methods that they use, and encourage them to adopt safer ones.

- The involvement of males in the birth event has had varying outcomes, and many programmes have shown a positive effect on males. Although initially women may be reluctant to have their husbands present during the delivery of a child, many have subsequently expressed appreciation after the birth, stating that men tend to be more understanding. Such innovations can be undertaken only with support from other members of the community or health workers.
- Generally, men put a lot of emphasis on the source of information, and it should be recognized that concerning women's reproductive health issues it is important that this source be respectable to them. Nevertheless, men should also be encouraged to trust and communicate with their partner.
- Above all, perhaps, we should try to be creative in getting our message across. For example, the Zimbabwe National Family Planning Council launched what turned out to be a very successful educational programme for men, motivating them to be involved in family planning. It included a serial drama broadcast on the radio that reached many men. One study subsequently estimated that joint decision-making in matters of spacing children increased from 25 per cent to 35 per cent of couples between 1988 and 1989 (Finger, 1992). Such results are not beyond our sights.

Family relations and gender roles: the example of North Africa

Laila Shukry El-Hamamsy*

Several factors determine patterns of family formation and the status of women in a society. One important determinant is, naturally, the legal framework that defines men's and women's rights and responsibilities. The law and its sanctions obviously influence behaviour directly; they also do so indirectly by bolstering or weakening a person's bargaining position within the extremely significant relationships that are involved in family life. Yet the letters of the law do not tell all. Laws can be broken or circumvented; those whose rights are denied or those who are victims of abuse often cannot, or dare not, resort to legal action to gain protection or justice. Moreover, many areas of private behaviour either are not or are only minimally controlled by law. This is particularly true of family relationships and gender roles, which are largely conditioned by the socio-cultural environment, the values and sentiments of the individuals involved. To see how things really are, one must look beyond the law.

* Professor Emeritus, Social Research Center, The American University in Cairo (Egypt).

Before analysing family formation and the status of women in North African countries, it must be emphasized that these countries do not necessarily represent one homogenous whole, even though all are Arabic-speaking, Islamic nations. They have some common concerns and problems and much cultural affinity, but they have had distinct histories and cultural influences. Today, they exhibit important demographic, social, economic and political differences. Population size and per capita GNP are but two examples. Egypt has 63.7 million inhabitants, or over one-third of the region's population. The Libyan Arab Jamahiriya, the smallest country of the region, has 5.4 million inhabitants. The respective populations of Algeria (27.8 million), Morocco (28.9 million) and the Sudan (29.0 million) are less than half of that of Egypt. Tunisia, with 9.2 million inhabitants, has a population less than Egypt's capital, Cairo, which has about 12 million (Population Reference Bureau, 1996*a*). Using per capita GNP as an indicator, the World Bank classifies Egypt and the Sudan as low-income countries; Algeria, Morocco, Tunisia as medium-low income countries; and the Libyan Arab Jamahiriya as a medium-upper income country (Toubia, 1994).

The picture is further complicated by the fact that, within each nation, no single pattern of family life and gender roles can be considered as characteristic of all ethnic groups. In this chapter an effort will be made, whenever the available data permit, to point out the existing variations at both regional and national levels.

Legal framework

The constitutions of all North African countries stress the equality of men and women and give them equal political rights, which means that they have the right to vote and to run for public office, as well as equal rights to education and employment. The Egyptian constitution, for example, affirms that 'the State shall ensure equal opportunities for all citizens', and, furthermore, that 'citizens have equal rights and duties without distinction because of sex, origin, language, religion or belief' (Zulfikar, 1995). The Libyan Arab Jamahiriya's *Green Book*, which can be considered as its constitution, states the following:

> the woman is a human being and the man is a human being. . . . Hence the woman and the man are clearly equal in their humanity. Any discrimination between them is a crying injustice that has no basis. The only difference between males and females . . . is biological; and this only means that they have different needs, such as women's need for special care during pregnancy and for special arrangements, if working, to attend to her nursing and other young children (Kaddafi, 1990).

Despite the emphasis placed on gender equality by the respective constitutions, most laws relating to personal status or the family do not completely adhere to the principle of equality. Many provisions are based on the legislators' interpretation of Islamic or 'Sharia' law. The fact is, there is no set of Sharia laws to which everyone agrees or conforms. For many centuries, divergent theological opinions have been advanced – even by the four classical ninth-century schools of Islamic jurisprudence – as to what laws and practices are consistent with Islamic doctrine. The lack of uniformity in family laws is also due to differences in the extent to which the theologians and law-makers of a particular country believe 'reasoning' (a basic concept in Islamic theology) can be exercised to modify Sharia laws in response to the evolution of society.

With reference to Egypt, one critic refers to custom as another source of influence on the judicial process: 'In the absence of an applicable text of law, the judge would have to refer to custom. . . . This is where the courts find basis to apply the practices of . . . trade in a commercial case or the prevailing traditions within the relevant social class of society in a family law dispute' (Zulficar, 1995). She goes on to add that, with the recent emergence of some extremist religious tendencies in parts of the Arab world, court rulings based on custom have, at times, reflected less liberal definitions of gender roles and rights.

Family formation and family law

Marriage choice

In all North African countries marriage is the only socially acceptable means of starting a family. Because of severe sanctions against women's sexual relationships outside marriage, cohabitation of unmarried couples and single-parent families headed by unwed mothers are rare. This seems to be the case even among North African immigrants in Europe (Feld et al., 1996). Out-of-wedlock pregnancy is considered to be a family calamity and a dishonour; as a result, the woman is likely to be subjected by her family to severe punishment and may be forced to undergo an unsafe abortion.

According to traditional Islamic law, a woman can marry the man of her choice. The marriage is recognized if attended by two witnesses and if publicly declared. Yet, in Morocco and the Sudan, the father/guardian's consent is mandatory. In Tunisia, the Libyan Arab Jamahiriya and Egypt a girl has the right to get married and to sign her own marriage contract. In Egypt, however, the father can request the courts to annul the marriage on the basis of incompatibility.

In reality, decisions concerning marriage often involve the family. As with much else in family life, patterns range from very liberal, where the girl has the final word in all decisions relating to education, work, marriage and reproduction, to very conservative, where the parents have the upper hand in all important decisions. The younger, less-educated and more economically dependent a girl is, the less her freedom of choice, even after marriage, when the control over her life passes on to her husband and mother-in-law.

Except for the Sudan, all North African countries have set minimum legal ages for marriage. For females, it is 16 years in

Egypt and Morocco, 17 years in Tunisia and 21 years in the Libyan Arab Jamahiriya. Except for Egypt, marriage below the legal age is possible only through the courts. In the Sudan a girl can get married the moment she reaches puberty, while a girl of 10 years of age can do so only through the courts.

Analysing the results of national household surveys conducted between 1986 and 1995, one researcher found intra-regional as well as intra-national differences in the age at which women marry (Farid, 1996). He reports the following median ages at first marriage for the 25–29 age-group, which is indicative of recent trends: 23.8 years in Morocco, 23.2 years in Tunisia, 21.9 years in Algeria, 19.9 years in Egypt and 19.5 years in the Sudan (Farid, 1996). Within individual countries, the same study found that the median age was lower among the older age-groups, and lower among the less-educated and rural women within the same age-groups. In 1992 the percentage of married women in the 15–19 age category in North African countries was as follows: Tunisia 4.3 per cent, Morocco 11.3 per cent, Egypt 15.3 per cent, Algeria 22.5 per cent, the Libyan Arab Jamahiriya 36.7 per cent and the Sudan 41 per cent (Toubia, 1994).

The incidence of early marriage may, in fact, be greater than national surveys reveal because of people's awareness of the legal minimum age. In a study of early marriage in two highly conservative Egyptian villages, it was possible through indirect questioning to obtain women's true marriage ages. The results indicated clearly that the fate of a girl who did not attend school was inevitably early marriage. Over 90 per cent of the illiterates had married as teenagers, and 47 per cent of these marriages had occurred before the legal age of 16 years. On the other hand, the majority of those with secondary or higher education had married at age 20 or older, and none at less than 16 years. The women who were interviewed, including the uneducated majority, overwhelmingly supported the education of girls and the postponement of a girl's marriage until at least age 20, when, as they said, she would be mature enough physically and mentally to carry family responsibilities. Yet, many of them had accepted early marriage as the only pragmatic option for themselves as uneducated girls. As some explained: 'I never went to school and I had no job, so why wait?' This study and many others have shown that education is the key to enhancing women's status and to empowering them to exercise their rights (El-Hamamsy, 1995).

A woman's age at marriage is, naturally, much higher in urban areas, where a larger percentage of women are educated. Even though education is a universal right guaranteed by most of the constitutions mentioned above, and even though in some countries, such as Egypt and Morocco, it is free and compulsory for all children in the early years, many girls in rural areas are nevertheless denied an education. This is because of rural traditions that consider the home as the natural place for a girl, where her reputation and honour can be protected and where she can learn to become a housewife, or because of poverty and the family's inability to afford the costs of schooling or to dispense with a girl's labour either as an unpaid family-worker or as a wage-earner. Governments are making special efforts to increase rural girls' access to education by establishing community or one-classroom schools in remote rural areas.

Planning the family

It has also been observed that, regardless of how young a bride is, she is subjected to much pressure by in-laws, parents and neighbours to become pregnant immediately after marriage (El-Hamamsy, 1995). The young bride is often reluctant to postpone her first pregnancy in order to prove that she is fertile, or to gain status within the family, or to avoid the threat of divorce or of sharing the husband with a second wife. While North African countries may not have the serious problem of pre-marital teenage pregnancy that other, more permissive societies have, those where early marriage is still common do have a problem of early teenage pregnancy within marriage, accompanied by all the associated health risks to mother and child.

Except for pro-natalist Libyan Arab Jamahiriya, the North African nations have fertility control policies. Tunisia has the lowest fertility rate and population-growth rate, while the Libyan Arab Jamahiriya and the Sudan have the highest. The total fertility rate for Tunisia is 3.4; Egypt, 3.6; Morocco, 4.0; Algeria, 4.3; the Sudan, 6.1; and the Libyan Arab Jamahiriya's, 6.4. The annual rate of increase in the total fertility rate is 1.7 per cent in Tunisia; in Egypt and Morocco it is 2.2 per cent; in Algeria, 2.4 per cent; in the Sudan, 3.0 per cent; and in the Libyan Arab Jamahiriya, 3.7 per cent (Population Reference Bureau, 1996*a*).

All countries offer maternity and family-planning services, and women in increasing numbers are using modern contraceptives instead of abortion, as methods of birth-control. The access to health services of women who are strictly confined to the home is naturally very limited. As reported to the present author by the health services in a remote rural area of Algeria, some men do not allow their women to visit the health centres for maternal or family-planning services, or even to take their children for vaccination or treatment. Fortunately, such cases represent a small minority in the region.

Polygamy and divorce

The man's right to marry up to four wives and to divorce at will are two provisions in family laws that clearly discriminate against women. Even though only a small percentage of men think of, or can afford to have, more than one wife, polygamy and a man's divorce rights can constitute a serious threat to a woman's sense of security, especially if she is economically dependent, and can greatly weaken her bargaining position in the event of serious disagreement or dispute. Although both provisions have always been part of traditional Sharia law, this has not prevented some nations from putting restrictions on their practice. Tunisia, which has some of the most egalitarian family laws in the region, prohibits polygamy altogether. In Morocco and the Libyan Arab Jamahiriya a man cannot marry a second wife unless he obtains the written approval of his first wife or through a court verdict. It is interesting that, in the Libyan Arab Jamahiriya, all court judges in family dispute cases are women. In Egypt a man can marry four wives but must inform his current wife, who can sue for divorce on the basis that she has suffered harm. In Morocco and the Libyan Arab Jamahiriya a man can marry a second wife only through the courts and he must inform his current wife as well as obtain the approval of the prospective wife. In the Sudan a man is permitted, without any restrictions, to marry up to four wives. Although under traditional Sharia law the woman has no automatic right to divorce, she can obtain this right if she includes it as a condition in the marriage contract. In Egypt, for example, women activists are trying to inform women about their legal rights and the procedures for obtaining them. Because of women's reluctance to begin a marriage by demanding the right to divorce at will, a new marriage contract form is proposed that would include the woman's right to complete her education and to work within or outside the country, which, if not met, would give the woman the right to divorce.

Divorce for an economically dependent woman may mean a future without any financial support, for unless she has young children, she is entitled to alimony for only a limited period of time. Although one of the incontestable rights of women, derived directly from Sharia law, is the right to possess wealth and to manage personal financial affairs independently from the spouse, this gives security only to those who have wealth. Poor women without means or jobs remain completely dependent on their husbands, who are obligated, by law and tradition, to carry the financial responsibilities within the family.

Custody of children

In all North African countries women have rights of custody over their children depending on the age of the child: up to age 7 for the boy and age 9 for the girl in the Sudan; up to ages 10 and 12 respectively in Egypt; up to age 16 for both boys and girls in Morocco; and up to age 18 for boys and until marriage for girls in the Libyan Arab Jamahiriya. Tunisia has no age limits.

In Tunisia and Morocco the children of a woman married to a foreigner can acquire her nationality. In Egypt only males can pass their Egyptian nationality on to their children. This is another one of the provisions which defenders of women's rights in the region are trying to change.

Women's groups in all of these countries are trying to bring about a revision of family laws in order to eliminate discrimination and enhance women's status. In some countries, such as Egypt and Morocco, they are strongly supported by the national leaders. The resistance to change by conservatives has, as can be expected, not disappeared entirely.

The right of choice

Ideally, the right of choice in relation to marriage, reproduction and health should be the prerogative of a couple. In North Africa there are many families where the young spouses make their own decisions concerning the timing and spacing of pregnancies, the desired number of children, the use and choice of contraceptives and the termination of pregnancy or of marriage. Nevertheless, as noted earlier, the influence of parents and in-laws can be substantial, particularly in the early years of marriage, and even within educated families, although it may take the more subtle form of persuasion rather than of direct pressure. The implications of such influence is highly significant and ethically problematic: in societies where the family occupies a central place in the life of the individual and represents the main source of security and well-being, it may be difficult and unrealistic to try to exclude the family from such decisions. Efforts to improve women's health and to assist them in planning their families may fail to achieve their targets if influential family members are ignored. Ethical principles such as the right to confidentiality of health and genetic information or the right to informed choice and consent in relation to medical interventions if advocated as the exclusive right of the individual, may not be readily accepted in North African societies, where the family carries a great deal of responsibility for the health and welfare of its members.

Violence against girls and women in Latin America

Genoveva Keyeux* and Silvina Ramos**

In Latin America, as in other regions of the world, there is no consensus on the definition of 'gender-based violence', either from a theoretical or from a political point of view. The authors of the present chapter support the definition formulated by Heise (1994*b*), that gender-based violence 'includes any act of force or coercion . . . gravely [jeopardizing] the life, body, psychological integrity or freedom of women' that is based on an imbalance of power due to differences of gender. This definition encompasses 'rape, battery, homicide, incest, psychological abuse, forced prostitution, trafficking of women, sexual harassment, genital mutilation and dowry-related murder' (1994*b*). It also makes it possible to define as gender-based violence those social relationships and institutions that maintain and reinforce the subordination of

* Director, Molecular Genetics Unit, Facultad de Medicina, Pontificia Universidad Javeriana, Santate de Bogotá, DC (Colombia).

** Senior Researcher, Health Economy and Society Department, Center for the Study of State and Society, Buenos Aires (Argentina). The authors express their gratitude to the following people, who assisted in the preparation of this text: Lucy Cardona, Marisol Dalmazzo, Gabriel Lago, Margarita Córdoba, Luz Rivera and Mariana Romero.

women, as well as certain medical practices common in Latin America, such as unnecessary Caesarean sections and enforced sterilization.

Support for this type of definition is found in the 'Convention of Bélem do Pará', one of the most important recent legal documents to be endorsed.[1] Building upon the spirit of the 'Declaration on the Elimination of Violence Against Women' (United Nations, 1993*b*), this political statement recognizes that violence against women breaches a broad range of human rights. It defines a responsibility held by states, recommending that specific actions to eliminate violence, both public and private, be undertaken. Most notably, 'it gives individuals and groups the right to file petitions before the Inter-American Court of Human Rights when States fail to carry out the responsibilities defined in the Convention' (Copelon, 1996). Thanks to these documents, and also to the work of women's groups in the region, violence against women in Latin America is increasingly being acknowledged, first, as one of the most extensive violations of women's human rights and, secondly, as a severe health problem. Moreover, it is being acknowledged that almost all women are affected, irrespective of social standing.

Historical, political and cultural factors are inevitably involved in explaining gender-based violence, causing accounts to vary between countries. It can nevertheless be argued that such violence has one general cause: the subordination of women. And in Latin America, despite historical, political and cultural differences, researchers have identified at least three specific factors that explain gender-based violence. First of all, there is a clear division between public and private domains that dissociates domestic affairs from public scrutiny. Secondly, it is noted that violence is

1. This convention was adopted with acclamation during the 24th regular session of the General Assembly of the Organization of American States, on 9 June 1994, in Bélem do Pará, Brazil. It has since been ratified by the Parliaments of almost every country in the region.

often accepted as legitimate behaviour when conflicts arise in relationships. And thirdly, accepted stereotypes portray males as powerful and controlling and females as caring and subordinated, and this shapes what has been referred to as the 'machismo-marianismo' cultural pattern (Stevens, 1973; Riquer et al., 1996).

If gender-based violence is to be ended in Latin America, strategies must be formulated to challenge and change these cultural values. Legal reform must be introduced and governments must institute policies to assist those who have been victims to come forward. This chapter will bring attention to some of the most pressing concerns in Latin America.

Abuse of girls

Child abuse of every type may be found worldwide, in developed and developing countries alike. In Colombia (Ruiz Gonzalez, 1995) it is estimated that 58 per cent of all children abused are girls and that fathers are often the predominant aggressors. Mothers who neglect their children are also responsible. Most forms of violence are repeated throughout childhood, with serious effects on the physical and mental health of the child and, later, the adult. There is alarming evidence that many cases go unreported.

Physical abuse

In many developing countries, domestic duties must be undertaken early in life by girls, who may be forced to carry out tasks beyond their capacity and strength. For their inadequacy, they are often verbally and physically punished. In cases of extreme physical abuse, girls in developing countries either flee from home to seek employment as maids in wealthy people's houses, thus exposing themselves to other dangers and more abuse, or end up on the street as beggars, thieves or prostitutes.

In the Colombian study quoted above, cases of physical abuse accounted for over 37 per cent of the total cases of child abuse. Girls suffered slightly less from this type of abuse than boys. Since the father was the most frequent aggressor, the resulting physical injuries were usually severe, requiring medical treatment and even hospitalization.

Psychological abuse

Psychological abuse is often difficult to account for objectively, since often the parties concerned have grown so used to the forms of behaviour involved that they do not recognize them as abuse. Nevertheless, in the Colombian study (Ruiz Gonzalez, 1995) cases of psychological abuse accounted for 20.5 per cent of the total number of cases and it was found that mothers tended to mistreat their daughters psychologically more than fathers. Obviously, it is hard to dissociate psychological abuse from physical abuse, but clear forms of psychological abuse do exist, such as harsh verbal insults.

Sexual abuse

Girls are more frquently sexually abused than boys. More than 18 per cent of the girls surveyed in the Colombian study were sexual-abuse victims, compared to 3.8 per cent of boys. All age-groups were found to be vulnerable, with the most frequent cases occurring in girls between 10–15 years. Some cases were observed in infants between birth and one year (Ruiz Gonzalez, 1995).

Sexual abuse takes many different forms and has varying degrees of severity. In developing countries where girls have very little opportunity open to them, they may feel they have no choice but to sell their bodies in a 'partnership' that seems to provide security and economic stability. In such cases, girls are often forced into sexual initiation at a very young age. Another form of sexual abuse is sex reciprocation, wherein a girl or woman agrees to sex in exchange for some occasional payment, gift or favour. This practice is often observed in houses where maids consent to have sex with the elder sons, or even the husbands, in exchange for some complement to their salary.

Prostitution, which is common in developing and industrialized countries alike, often involves severe sexual abuse. Furthermore, it often involves young girls and adolescents. Although governments do possess information about this problem, little is being done and the situation is getting worse. In Bogotá, for example, the numbers of very young girls and adolescents involved in prostitution has doubled in three years, with over 4,000 girls implicated (Camara de Comercio de Bogotá, 1994; Defensoria del Pueblo, 1995*b*). And in Colombia over 48 per cent of adolescent prostitutes are aged between 14–16 years, while in some regions of the country more than 10 per cent of them are as young as 10 years old. Most of them abandon school, although for some, prostitution is a weekend occupation (Defensoria del Pueblo, 1995*b*). The consequences of such early and frequent sexual activity may include sterility, caused by inflammatory pelvic disease, alcoholism and drug-addiction.

Prostitution is a lucrative international business, and Latin America and Southeast Asia are preferred 'sources' of young prostitutes. Girls from these regions often end up in the brothels of industrialized countries, where they become literally slaves of the proprietor. They are commonly driven there by precarious living conditions in their own countries, and it is not unusual for them to be lured by pimps. Other factors are also apparent. One recent study found that 35 per cent of Colombian prostitutes had turned to prostitution because of family disintegration, 19.3 per cent because of physical abuse and 20.8 per cent because of psychological abuse (Defensoria del Pueblo, 1995*b*).

Violence in the schoolroom

Physical, psychological and even sexual abuse also frequently occur in schools, where girls are subject to mistreatment from teachers. The teachers may promise good reports in exchange for sexual favours, and if pregnancy results the adolescent is expelled. In such cases guilt falls solely on the girl, with blame also coming from schoolmates and her own family. Even in higher educational institutions, professors obtain sexual favours in return for better grades (Londoño, 1991). Invariably, the educational opportunity is destroyed, along with the victim's psychological balance.

Neglect

Neglect is characterized by situations in which basic material and affective needs, such as food, education, care, etc., are intentionally withheld. In general, neglect is more common when the parents are young and is more frequently inflicted by the mother. In Colombia, 14 per cent of the cases of child abuse were characterized as neglect, and girls and boys suffered more or less equally (Gonzalez, 1995). One of the most serious forms of neglect is total abandonment, and the children who suffer from this often drop out of school and find themselves living on the street.

Violence against women

During the past two decades women's groups have begun devoting systematic efforts to address the general issue of violence against women. These groups[2] have focused on developing

2. In 1990, ISIS Internacional published a directory identifying some four hundred independent women's organizations working on issues related to gender violence in Latin America. In 1992 the Latin American and Caribbean Feminist Network against Domestic and Sexual Violence was created with members from twenty-one countries.

resources and strategies to help those women affected: programmes of psychological and legal counselling for the battered and raped; centres for the victims of domestic violence; effective means of heightening public awareness; and the promotion of research and training activities for concerned professionals, such as police officers, lawyers, judges, physicians, journalists and government officials.

Domestic violence

Domestic violence includes any type of deliberate physical, verbal or psychological aggression against a family member that causes emotional or bodily harm. For various reasons, comprehensive statistics are hard to establish. It is estimated that many cases go unreported, and the reasons are complex: women are often educated to endure such abuse; they are often economically dependent; and legislation in many countries does not define the more subtle forms of domestic violence. In many Civil Codes, for example, domestic violence is considered as 'Person's Lesions', wherein visible damage or testimony from third parties is required for a trial.

Nevertheless, the available statistics about domestic violence are alarming. In Santiago de Chile, 60 per cent of women in one survey reported having been abused, and 26 per cent reported having been physically abused by their partner (Larraín, 1993). In studies conducted in Colombia, 20 per cent of the urban and rural women interviewed claimed to have been physically abused, 33 per cent psychologically abused and 10 per cent raped by the husband or partner (Profamilia, 1990; Londoño, 1991; Departamento Nacional de Planeación, 1995). In Mexico, statistics reveal that rape occurs 125 times per 100,000 inhabitants (Shrader-Cox, 1992). In Quito, Ecuador, 60 per cent of women from the lower-income bracket reported having been 'beaten' by a partner. Finally, in Buenos Aires, Argentina, 78 per cent of domestic accusations presented before the public services in

1993 were caused by injuries inflicted by the woman's husband or partner (Valdes and Gomariz, 1993).

Rape

Rape again presents a situation that is hard to quantify objectively. It is well known that in many cases which do come to light the woman herself is blamed for having incited the man to act. Situations in which women are forced by a partner to have sexual intercourse are only now being recognized as instances of rape. In many countries, even if rape results in pregnancy, abortion is considered illegal and is punished just as severely as other instances of abortion. For these and other reasons, it is estimated that only 70 per cent of cases are actually reported (Londoño, 1991; Defensoria del Pueblo, 1995*a*).

No particular social type has been attributed to the rapist. It has been noted that guilty males may be of any socio-economic level and cannot necessarily be identified as 'deranged'. In the United States it is estimated that one women is raped every six minutes and, furthermore, that 15 per cent of these rapes involve the husband (Londoño, 1991). Some cultural features seem to play an important role. For example, *droit du seigneur* is thought to account for the rape of many youngsters by their fathers, grandfathers, uncles or other close male relatives.

Sexual harassment in the working environment

The cultural attitudes that give rise to abuse in the home are partly responsible for sexual harassment in the workplace. Again, women do not report all such occurrences. They may fear demotion or dismissal, or they may fear being blamed, mocked or humiliated by their colleagues.

Institutional violence: family planning and abortion

It can be argued that barriers to family planning and legal abortion constitute an example of ongoing institutional violence in disregard of women's rights. Family planning in many developing countries is still a marginal policy, and women are often obstructed from choosing to limit the number of children they will have. In Colombia, poor women still have the same number of children as their mothers' generation, although the global fertility rate has dropped (Libro Blanco de la Mujer, 1994).

In Latin America, with the exception of Cuba and Puerto Rico, abortion is a crime. Penal codes discipline not only the women who undergo abortions, but also those who perform them. Where abortion even for therapeutic reasons is prohibited, women do not benefit from prenatal diagnosis, and when there are good reasons to suspect that a foetus is malformed or severely ill, the women must continue her pregnancy or resort to clandestine abortions. For legal restrictions do not stop abortions, and, as in other parts of the world, such restrictions only create the possibility that women will suffer the physical and psychological consequences of unregulated and unsafe operations.

A recent study carried out in six countries estimated the annual number of abortions in the region of Latin America at four million (Henshaw, 1990). Although official maternal mortality rates have been declining over the last decade in almost all of the countries, complications resulting from unsafe abortions are still the primary cause of maternal deaths and the second leading cause of women's hospitalizations after delivery (FLACSO, 1995).[3] The estimated annual mortality rate due to

3. It is very important to note that studies in several countries have shown that official maternal mortality rates do not reflect the actual prevalence of deaths owing to a serious problem of under-reporting (WHO, 1994*a*).

unsafe abortions is 58 per 100,000 live births in South America and 23 per 100,000 in Central America (WHO, 1994a). The illegality of abortion also has a direct impact on the quality of health services given to women suffering from complications due to clandestine abortions. Various studies have shown that women in these situations often suffer abuse of several types: disinfecting wounds without anaesthesia, psychological intimidation, humiliation and threats of police prosecution (Ramos y Viladrich, 1994).

Case Study

Women as victims of war

Berit Schei, Amira Frljak, Mihr Pjskic, Monika Hauser*

In the autumn of 1995 a fragile peace agreement was signed after four years of war in Bosnia and Herzegovina. During the war people endured great suffering: civilian areas were deliberately attacked; people were expelled from their homes and were forced to flee, saving nothing but their lives; and women, children and men were kept in concentration camps or in detention, where many were mistreated, tortured or murdered. The United Nations High Commissioner for Refugees (UNHCR) has estimated that about half of the population of Bosnia and Herzegovina was displaced, and although many people sought refuge in other countries, the majority remained in Bosnia and Herzegovina and continued to suffer from the effects of the war (UNHCR, 1995).

* B. Schei is at the Norwegian University of Science and Technology, Trondheim (Norway). A. Frljak and M. Pjskic are at the Medical Women's Therapy Centre, Zenica (Bosnia and Herzegovina). M. Hauser is at Medica Mondiale, Cologne (Germany).

Rape during wartime

One of the atrocities of the war in the former Yugoslavia was systematic rape, a pattern of behaviour that was identified in a report from a United Nations commission of experts (United Nations, 1994a). The findings of the report were based on interviews conducted in March 1994 with victims and witnesses from Bosnia and Herzegovina who had fled to Croatia or Slovenia. Although known to have occurred in many wars (Brownmiller, 1975), rape is an issue that has been grossly neglected by the international community. Only afterwards, and mainly because of the media focus on rape in the former Yugoslavia, was there any increase in public awareness (Swiss and Giller, 1993; Acheson, 1993; United Nations, 1993a).

In all, five different patterns of behaviour were identified. The first pattern included rape committed by individuals or small groups in conjunction with looting. These incidents took place before any widespread fighting broke out in the region. The second pattern involved individuals or small groups who raped while fighting in the area. When forces attacked towns or villages, the population was rounded up and divided by sex and age. Some women were raped immediately in their homes, others were selected after being grouped and then raped publicly. The third pattern of rape occurred when women were taken into detention. They were either kept in temporary detention in the villages or towns or taken to camps further away. Soldiers, guards, paramilitaries and even civilians were allowed to enter the camps, pick out women, take them away and rape them. Beating and torture occurred as part of the sexual assaults. After such incidents the women were either killed or taken back to the villages, towns or camps they had come from. The fourth pattern of rape occurred for the purpose of terrorizing and humiliating women as part of 'ethnic cleansing'. To impregnate the women was part of the pattern, and pregnant women were detained until it was too late to obtain an abortion. The fifth pattern of rape involved the detention of women in hotels for the sole purpose of providing sexual distraction for soldiers.

Services to help raped and traumatized women were set up in Croatia. In Zagreb, the Rehabilitation Centre for Torture Victims (RTC) played an important role in this work (Arcel et al., 1995). However, for women trapped inside Bosnia and Herzegovina, the situation was more difficult. In what follows we will report from two projects set up inside the war zone in Zenica to meet the needs of women and children.

The women's therapy centre in Zenica

Zenica is an industrial town with 120,000 inhabitants. It was frequently shelled, and war activities in the nearby areas could often be heard. From the spring of 1993 and throughout a whole year, the area suffered severe difficulties due to blockades established by the so-called Bosnian/Croatian defence council. There was a lack of all kinds of essential supplies: food, petrol, diesel and medical supplies. Zenica received some 50,000 refugees during the war, mostly women and children. Most men between the ages of 18 and 60 years were either fighting, captured, wounded, dead or missing.

The majority of the refugees arriving in Zenica were given accommodation by relatives. However, many had to be sheltered in schools and sports halls. The living conditions in these places were unequivocally appalling. The sanitary facilities were often destroyed. There was a lack of water and electricity. All refugees were totally dependent on humanitarian aid. In this desperate situation the public facilities were reduced to an absolute minimum.

In response to the demand for services for women, one of the authors of this case study went to Zenica at the end of 1992. Together with the local professionals, she established a Women's Therapy Centre, which opened in April 1993. The Centre was based on the understanding that women needed a range of integrated services. These included the following: (1) an out-patient gynaecological clinic with an operating theatre and a mobile team with a vehicle serving as an ambulance carrying gynaecological equipment; (2) general medical service provided by a general practitioner; (3) psycho-social services; and (4) shelters and educational programmes. Refugee women came to the Centre with a range of problems and were given support and treatment according to their needs and the resources available. Rape commonly results in severe and long-lasting psychological problems (Dahl, 1993). Some of the most traumatized women were given the opportunity to live in the established shelters. There they could both communicate with other women and have access to professional assistance. Many of these women had been raped, and approximately 30 per cent of those women who had been raped were pregnant as a result. In most cases the women reached the Centre too late to have an abortion and so they eventually gave birth. The patterns of rape reported by these women corroborated the findings of the United Nations report described above (1994*a*). The women were given psycho-social support, medical treatment and psychotherapy, in addition to formal education in order to restore their psychological and physical health.

Women seeking gynaecological services addressed themselves directly to the clinic. The procedures for gynaecological examination and treatment were all adapted to the 'as if raped' approach. This meant that care was taken to reduce the potential danger of triggering a traumatic 'flashback' experience. An ambulance visiting the surrounding villages was equipped with gynaecological examination facilities.

Approximately 25,000 gynaecological out-patient consultations were performed between 1993, when the Centre was opened, and early 1998, when this text was prepared. The Centre conducted a study based on information from 486 records of consecutive gynaecological consultations in four different periods between 1993 and 1994, and in 3 per cent of the records obtained from refugee women, this work has revealed a pattern of rape not previously identified (Frljak et al., 1997). In the gynaecological consultations, no attempt was made to ask routinely about a history of rape. Thus, many women who were raped may have received gynaecological care without disclosing this information.

Women as victims of war trauma

Another non-governmental humanitarian organization, Norwegian People's Aid (NPA), established services for women and children in Zenica. Women living together with their children in some of the refugee accommodations were invited to attend a new women's centre that opened in September 1993, offering occupational activities in groups, structured conversations in a group setting, educational and recreational activities and opportunities to receive counselling. The invitation was open and the women who attended could thus be considered representative of the refugee women in Zenica. A study based on a questionnaire was conducted among 209 of these women in 1994. Information on war-related traumatic events, socio-demographic factors and post-traumatic stress symptoms was collected. Of these random women, 26 per cent had been detained, 11 per cent had been in concentration camps and 22 per cent had witnessed killing. Nearly all had felt their life endangered. Most of the women had witnessed the destruction of their homes. Suffering from post-traumatic stress symptoms was common, in particular among women who had survived the most severe traumas: concentration camps or other kinds of detention (71 per cent) (Dahl et al., 1998). This demonstrates an extremely high level of trauma among the displaced women within Bosnia and Herzegovina during the war and the need for continued support.

Conclusion

The Vienna Declaration stated that violations of 'the human rights of women in situations of armed conflicts are violations of the fundamental principles of the international human rights and humanitarian law' (United Nations, 1995*b*). The atrocities women suffered during the war in Bosnia and Herzegovina demonstrate obvious violations of women's human rights. The establishment of reproductive services and psycho-social assistance for women in emergency situations has become an important task for the international community. The issue was recently addressed at a meeting of the World Health Organization, and the conclusion was that sexual abuse and rape during war must be addressed as a particular reproductive-health issue (Turmen, 1995).

Case study

Sex traffickingand prostitution: human rights and health consequences

Janice G. Raymond and H. Patricia Hynes*

Trafficking in women and children is at the same time so widespread, yet so invisible. Its invisibility is anchored in two kinds of foundations: the traditional view of gender inequality which instrumentalizes women's bodies for sexual and reproductive use; and the more liberal view which redefines some forms of sexual exploitation such as prostitution as work, legitimates the selling of sexual 'services' as commerce, and reconstructs the female body as a commodity.[1]

The United Nations estimates that 4 million women and children are trafficked worldwide.[2] The most rampant forms of trafficking are for sexual exploitation, particularly for prostitution, mail-order bride industries and sex tourism. Women and children are also trafficked for bonded labour and domestic work, and much of this trafficking concludes with women being sexually exploited as well.

---

* Coalition Against Trafficking in Women (United States).

1. The position that prostitution should be redefined as work, that women in the sex industry sell 'sexual services', and that the sex industry should be recognized as a legitimate economic sector is gaining support among certain organizations. See, for example, the 1998 report of the International Labor Organization, Lim, Lin Lean (ed.), *The Sex Sector: The Economic and Social Bases of Prostitution in Southeast Asia*, Geneva, International Labor Office, 1998.
2. United Nations estimate cited in Michael Specter, 1998, 'Traffickers' New Cargo: Naive Slavic Women', *New York Times*, January 11, p. 6.

Defining the problem

Currently, there is an international debate around the definition of trafficking and whether to separate trafficking from prostitution. We support the definition of trafficking that has been defined by a network of international organizations as 'the recruitment, transportation within or across borders, purchase, sale, transfer, receipt or harboring of a person for the purpose of prostitution or exploiting the marriage of such person'.[3] Exploitation, instead of coercion, is the 'operative' concept in this definition. Other definitions have focused on consent, but the 1949 United Nations Convention for the Suppression of the Traffic in Persons and of the Exploitation of Prostitution of Others and Article 6 of the United Nations Convention on the Elimination of All Forms of Discrimination against Women are representative of a consensus in international law, that human trafficking is the recruitment and transport of persons for the purpose of sexual exploitation, regardless of whether or not they have 'consented' to their trafficking.

Countries as diverse as Viet Nam, Cuba and those in eastern Europe and the former Soviet Union – all in the process of becoming market economies, and concommitantly beset by enormous

---

3. This definition, and the rationale for framing the definition in this way, are contained in a letter to United States Senator Paul Wellstone written and signed by a number of individuals and women's rights organizations united against trafficking in women and children. The network of individuals and organizations – including Equality Now, Planned Parenthood Federation of America, International Women's Health Coalition, NOW, WEDO, Catholics for a Free Choice, the Protection Project, Coalition Against Trafficking in Women, the Sisterhood is Global Institute, National Black Women's Health Project, the Feminist Majority, Gloria Steinem, and the Center for Women Policy Studies – cites the need for changes in proposed U.S. Legislation on trafficking (S.600) that Wellstone has introduced into the United States Senate. For the text of this definition and letter, see the website of the Coalition Against Trafficking in Women at www.uri.edu/artsci/wms/hughes/catw

financial crises – are witnessing an alarming increase in trafficking and prostitution. Mail-order bride industries particularly capitalize on the trafficking of Russian women, as well as Asian and Latin American women, to men in industrialized countries who want foreign wives whom they consider to be pliable and exotic.[4]

In the Asian region alone, from 200 to 400 Bangladeshi women are illegally transported into Pakistan monthly; from 7,000 to 12,000 Nepali women and girls are sold yearly into the brothels of India; and 150,000 Thai and Filipino women are sexually trafficked yearly to Japan (see Fig. 1 and accompanying texts, pp. 126–9).[5] The trafficking of girls from Nepal to India is probably the most active sexual slave trade anywhere in the world. In Asia, millions of women and children have been led into systems of prostitution such as street prostitution, sex entertainment clubs, sex tourism, and brothels that may literally be cages or, conversely, luxury establishments. Brothels in Bombay and Delhi

---

4. Postings on the Internet promoting prostitution, pornography, sex tourism and mail-order brides around the world can be found at numerous locations and are expanding exponentially. The oldest forum on the Internet for promoting sexual exploitation is alt.sex.prostitution whose aim is 'to create market transparency for sex-related services'. Postings are then archived into a worldwide web site called *The World Sex Guide*. For a documentation and critique of these sites, see Donna M. Hugues, 'Protecting the Dignity of Women: Policing the Internet', *Report of the Proceedings of the First Major International Conference on Combating Pornography and Violence on the Internet*, pp. 13–17, London, Association of London Government, February 13–14, 1997.
5. The Bangladeshi-Pakistan and Nepal-India estimates come from the Centre for Women and Children Studies, p. 33, *Fact-Finding National Workshop on Trafficking in Women and Children*, Dhaka, Bangladesh, May 23–25, 1997. However, a report from the Lawyers for Human Rights and Legal Aid (LHRLA), *Trafficking of Women and Children in Pakistan: The Flesh Trade Report* 1995/96 puts the figure higher, at 100–150 Bangladeshi women per day trafficked into Pakistan. The Philippines and Thai to Japan statistics come from Coalition Against Trafficking in Women, Asia Pacific Map Project, *Trafficking of Women for Prostitution in the Asia Pacific*, CATW–Asia Pacific, Manila.

receive trafficked women from Bangladesh and are often the transit point for moving women to Europe and North America.

Causal factors

What are some of the reasons for the expansion of the global trade in women and children? Military presence, cultural practices, economic development that increases women's poverty and loss of power, ideologies of sexual liberalism and sexual conservatism, and an organized recruitment of women and girls by a multinational sex industry often assisted by governmental inaction and collusion with these sex industries all play a part in this expansion of the trafficking and prostitution of women and girls. Even United Nations peacekeeping forces in the former Yugoslavia, Kampuchea and Mozambique have organized prostitution rings while on United Nations peacekeeping missions.[6]

Globalization of the world economy also means globalization of the sex industry. Sexual exploitation moves freely across local and national borders. The economic development policies of many countries are locked into repaying foreign debt and loan

---

6. There is ample documentation on the use of women and children in prostitution by United Nations peacekeeping troops in various parts of the world. See, for example, Stanley Meisner, 'Prostitution Report Accuses United Nations Troops in Mozambique', p. A11, *Los Angeles Times*, February 26, for an account of United Nations troops in Mozambique who used young girls as prostitutes. The 22,000 UNTAC troops in Kampuchea were reported to have created a prostitution industry there by recruiting Vietnamese women by the truckloads and bringing them across the border into Cambodia. See Aurora Javate de Dios, 'The Global Trafficking in Asian Women', *Women Empowering Women: Proceedings of the Human Rights Conference on the Trafficking of Asian Women*, p. 16, Quezon City, Philippines, 1993. And the most well known of these examples is the sexual assault of women in the rape camps in Bosnia who were treated as prostitutes by troops of the United Nations Protective Forces (UNPROFOR). Reported quite widely; see for example, Roy Gutman, 'Bosnia Betrayed', *New York Newsday*, November 1, 1993, pp. 7 and 26.

Figure 1. Trafficking in women for prostitution in the Asia Pacific region (see

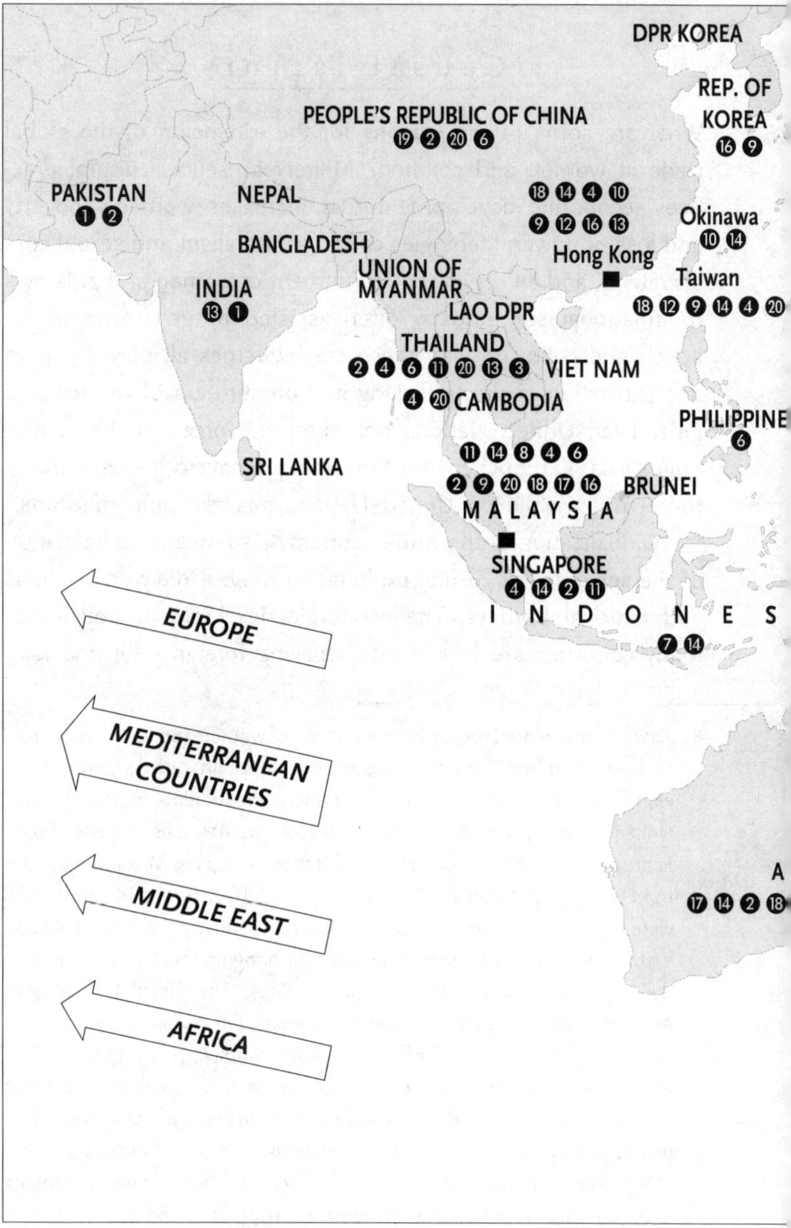

Source: The coalition against trafficking in women, 1996. Adapted from the map (and text) fir

* Official names: Hong Kong Special Administrative Region of the People's Republic of China

accompanying texts, next two pages)

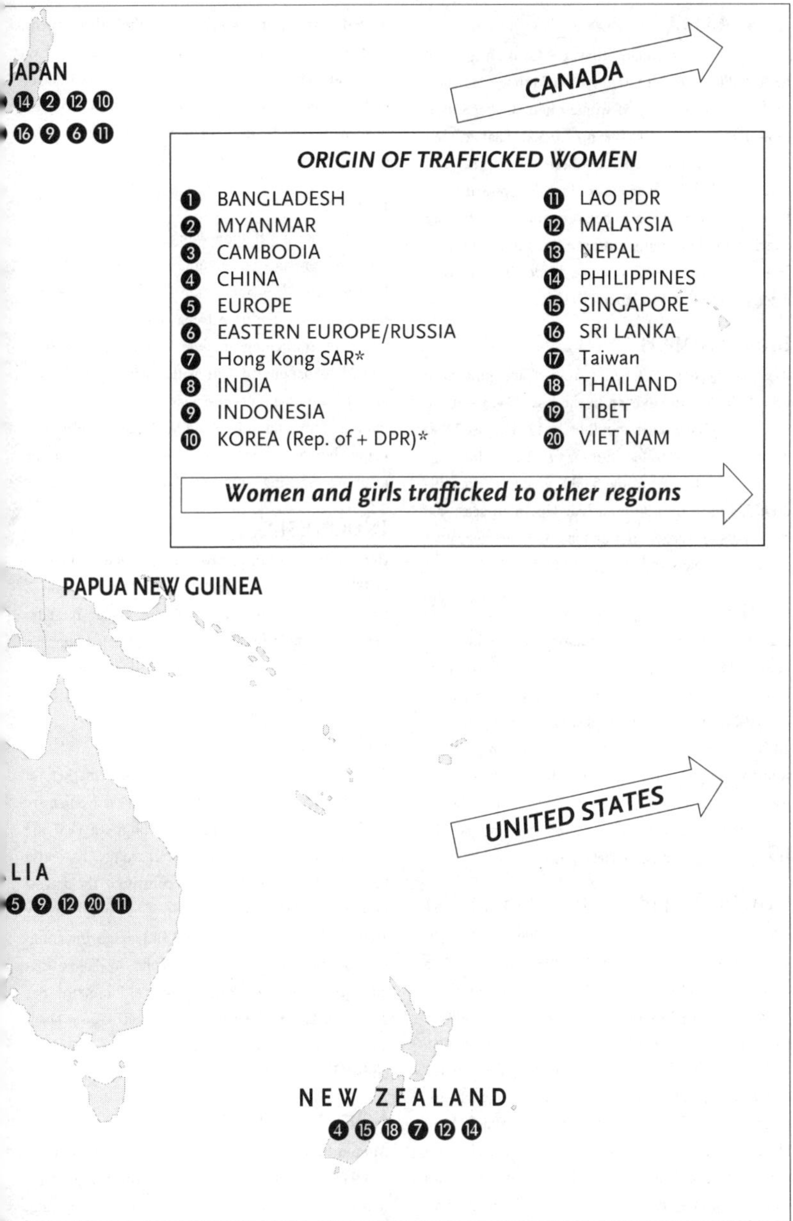

Democratic People's Republic of Korea and Republic of Korea; Lao People's Democratic Republic.

Trafficking in Women fo

AUSTRALIA

Federal police estimate that prostitution grosses A$30 million annually. International crime syndicates traffic in both drugs and women; ten smaller syndicates are known to traffic up to 300 Thai women yearly. Particularly in Canberra, Victoria and Queensland, Asian women are found in prostitution. Russian women have also been recruited for 'tabletop dancing' in clubs that often have links to brothels. In 1995, the Australian Council of Trade Unions recognized prostitution as a labour sector.

BANGLADESH

It is estimated that 200,000 women and girls have been trafficked to Pakistan in the last 10 years at the rate of 200–400 women monthly. In 1994 alone, 2,000 women were prostituted in 6 major cities in India. In Dhaka, almost 2,000 of the 5,000 prostitutes are children. Forms of trafficking include false marriages, sale by parents to 'uncles' offering jobs, auctions to brothel owners or farmers and outright abduction.

CAMBODIA

Current estimates put the number of prostitutes at 10,000–15,000, with 35 per cent minors. The figures were about 6,000 in 1991, but after the arrival of the United Nations UNTAC troops, the numbers rose to 20,000 in 1992. Almost half (48 per cent) of the women and girls in brothels had been abducted and sold; some were resold to other brothels or to traders who then smuggled them out of the country, for example to Thailand and Viet Nam.

CHINA (including Hong Kong SAR)

There is a resurgence of prostitution and trafficking in women and girls all over China, involving a high percentage of children and minor adolescents. In 1994, 15,000 cases of women sold as wives or for prostitution were handled by the police. For 1993/94, the figures released by the INFLS indicated that 24,751 women and 2,731 children had been rescued. In some regions, Vietnamese and Tibetan women and women from Myanmar have also been trafficked. Shangchuando island (off Guandong) is a tourist spot offering drugs and sex casinos, with 3,000 women from all over China. In Hong Kong SAR, fake contracts, often for domestic work, land women in brothels that employ Chinese guards to prevent runaways. An influx of East European women in high-priced clubs has been noted, and Russian Mafia members are said to be bringing women to Macau. In 1994, a women attempting to escape a sex establishment was murdered.

INDIA

An estimated 2.3 million women are in prostitution (of which one quarter are minors) in over 1,000 red-light districts throughout India. Cage prostitutes are often minors, often from Nepal and Bangladesh. The forms of trafficking include economic incentives offered to parents to part with their children, fake jobs or marriage promises, abductions. The promotion of tourism in Goa and Madurai, two of India's major beach holiday destinations, appears to be resulting in rising numbers of prostituted children.

INDONESIA

Here 71,281 prostitutes have been registered, of whom 60,000 are between 15 and 20 years of age. Localized bordello complexes, 'localisasi', are managed under local government regulations. The estimated financial turnover of the sex industry ranges from US$1.2 billion to $3.6 billion.

JAPAN

This country is the largest sex-industry market for Asian women. Over 150,000 non-Japanese women are working in prostitution, most of whom are Thai and Filipino women. Eastern European women have also been observed. Japanese men constitute the largest number of Asian sex tourists. The sex industry accounts for 1 per cent of GNP and equals the country's defence budget. One 'sex zone' in Tokyo that measures less than half a square kilometre has 3,500 sex 'facilities': strip theatres, peep shows, 'soaplands', 'lovers' banks', porno shops, sex telephone clubs, karaoke bars, private clubs and the like.

KOREA (REP. OF + DPR)

Around the military bases, there are 18,000 registered and 9,000 unregistered prostitutes. Forms of prostitution include escort and call girls, street prostitution, contacts in cafes, clubs, cabarets, show cases, massage

* Adapted from the map (and text) first established by the Coalition Against Trafficking in Wom

'rostitution in the Asia Pacific*

parlours and beauty shops. Women suspected of prostitution can be confined in rehabilitation centres without due process.

MALAYSIA

An estimated 142,000 women are in prostitution; between 8,000–10,000 are in Kuala Lumpur. The main channels are the 'recreation business', i.e. entertainment, fitness clubs and the like.

MYANMAR

An estimated 20,000–30,000 prostituted women and girls from Myanmar are to be found in Thailand; about 1,000 from the Shan state are in Chiang Mai. The forms taken by trafficking are false job placements that send women to brothels; abduction by agents for clients; sale of girls from hill tribes. Because they are illegal immigrants in Thailand, the prostitutes are arrested, detained and deported back to Myanmar; 50–70 per cent are HIV positive.

NEPAL

An estimated 5,000 women and girls are trafficked to India from this country yearly. Hong Kong SAR is the second biggest market after India. Brokers buy girls, especially in rural areas, and even family members sell girls to brothels (husbands sometimes even sell their own wives).

NEW ZEALAND

There are 6,000–8,000 prostituted women here. In Auckland, 800 out of 4,000 prostitutes are Thai, and there are 400 other Asian women. The channels include false employment offers; sponsorship by boyfriends or fiancés for residency; and 'debt bondage', used to keep women in prostitution. New Zealand is also used by traffickers in Thai women as a departure point for Japan, Australia and Cyprus.

PHILIPPINES

There are an estimated 300,000 women in prostitution, with 75,000 prostituted children. 'Entertainment' is the main channel, but a range of establishments from dirt-floor beer houses to karaoke clubs to beach resorts to expensive health clubs provide prostitution for men of every social class. Government policies favour the export of entertainers and domestic helpers, and this encourages the sexual exploitation of women. Further, government approval of 'R and R' (rest and relaxation) privileges for the United States Navy sustains the infrastructure of military prostitution as a system.

SRI LANKA

It is estimated that there are 15,000 prostitutes in the streets and in licensed and unlicensed massage parlours and brothels, including 30,000 prostituted children. 80 per cent of the labour migration in 1994 was women 'workers'. Job trainees in Korea and Japan have disappeared into underground labour markets that include prostitution.

TAIWAN

There are an estimated 40,000–60,000 prostituted children. Forty per cent of young prostitutes in the main red light district are Aboriginal girls. Girls under 13 have been made to undergo hormone injections by brothel owners to hasten their physical development. About 70 per cent of 1,771 women from Thailand, detained between 1992–1995 for illegal sojourn in Taiwan, were in the 'entertainment' sector.

THAILAND

Estimates of women in prostitution range from 300,000 to 2.8 million, of which one third are minor adolescents and children. Thai women are also in prostitution in many countries in Asia, Australia, Europe and the United States; some 4.6 million Thai men and 500,000 foreign tourists regularly use prostituted women and girls on a regular basis.

VIET NAM

It is estimated that between 60,000 and 200,000 women and girls are in prostitution, with 6.3 per cent under the age of 16. Women and girls are kidnapped for brothels, deceptive offers are made for jobs or tourist trips, or marriage is arranged with foreigners who sell and resell the women abroad. Organized tours of Taiwanese men come to buy brides for US$3,000.

obligations to industrialized countries who, in many parts of the world, have freely plundered the resources of those very nations that are in debt bondage. Many countries encourage their citizens to leave the country for work so that the payments, which workers send back to families, can stimulate and stabilize the economy. The Philippines, for example, has been forced into economic development policies based on exporting large numbers of their unemployed population overseas. Sixty percent of all overseas contract workers leaving the Philippines are now women;[7] and in 1992, 65 per cent of Filipino female migrants to Japan were 'entertainers', a code word for 'prostitutes'.[8]

In countries such as Thailand and also in some European countries, prostitution and sex entertainment are calculated and institutionalized parts of the economy, making use of the media, airlines, hotel chains, international communications and travel agencies and banks. The sex industry is so pervasive in Thailand that it accounts for 14 per cent of the gross national product (GNP).[9] This means that the sex industry exists and develops with the tacit or overt agreement of governmental institutions and is, in fact, part of their economic development policy.

To view trafficking for prostitution only through an economic lens, however, is to omit other key considerations. The economic reasons why individual women enter prostitution and/or are trafficked are fairly straightforward. What is less obvious is the gender bias for the facts that prostitution or sex entertainment are often the only forms of work that allow women to earn enough money

7. Philippines National Statistical Office, Survey of Overseas Workers, 1994.
8. Women's Education, Development, Productivity and Research Organization (WEDPRO), Inc., *Halfway Through the Circle: The Lives of Eight Filipino Women Survivors of Prostitution and Trafficking*, p. 8, Manila, 1998.
9. S. Tunsarawuth, 'Thailand's $44 Billion Underworld Prostitution Heads List of Lucrative Illegal Businesses, Says Study', *Straits Times* (Kuala Lumpur), 3 December 1996.

to survive; that prostitution is so overwhelmingly the selling of women's and girls' bodies to men; and that poor men – no matter how poor they are – can somehow always afford to buy poor women for the sex of prostitution. An economic analysis leaves unaddressed a supposedly 'natural law' that men's inherent needs must be satisfied, and that therefore prostitution is inevitable.

It has also become quite common for poverty to be invoked as an explanation for the failure of governments to enforce laws against trafficking and prostitution. Vitit Muntarbhorn, former United Nations Special Rapporteur on the sale of children, child prostitution, and child pornography, remarked, 'the poverty argument has been overplayed'.[10] It is not simply poverty but a poverty that is preyed upon by procurers and traffickers. A Thai NGO maintains that it is the presence of recruiters in villages in northern Thailand that governs the rate at which girls are sold into prostitution. In poor villages where agents and recruiters have been prohibited from operating by the local elders or officials, there is no or little selling of girls into prostitution.[11]

Most trafficked and prostituted women and girls experience extremely oppressive and exploitative conditions, brutality from pimps and recruiters, physical and sexual assault from the men who buy them for sex, and the same injuries that women who are battered in domestic relationships suffer.[12] The difference is that in prostitution everything that happens to women is called 'sex', not 'violence against women'. Increasingly, prostitution is being exempted from the category of violence against women.[13]

---

10. Cited in Cameron W. Barr, 'Getting Adults to Think in New Ways', *Christian Science Monitor*, 16 September 1996, p. 9.
11. Ibid., p. 11.
12. Janice G. Raymond, *The Health Effects of Prostitution*, North Amherst, Coalition Against Trafficking in Women, 1998. Available at www.uri.edu/artsci/wms/hughes/catw
13. Janice G. Raymond, 'Prostitution as Violence Against Women: NGO Stonewalling in Beijing and Elsewhere', *Women's Studies International Forum*, Vol. 21, No. 1, 1998, pp. 1–9.

Health consequences

Women in prostitution suffer, in magnified and epidemic proportions, the multiple health crises of women who are victimized by other forms of male violence and sexual exploitation. These include bodily injury and death, severe stress and trauma, a host of sexually transmitted diseases, infections and non-infectious diseases, unwanted pregnancy, multiple abortions, infertility and the risk of cervical cancer. Yet, historically, the overriding preoccupation of medical and public health practitioners regarding prostitution has been prostitutes as vectors of disease.[14]

The physical violence and trauma that prostituted women suffer at the hands of pimps and purchasers have largely been ignored, undocumented, and unaddressed by the health professions. Only recently, as survivors of prostitution are organizing alternatives for women in prostitution, are studies being undertaken to document the health impacts on women and girls in the sex industry. A 1994 study of 68 women in Minneapolis/St. Paul in the state of Minnesota (United States), conducted with women who had been in prostitution for at least six months, found that half the women had been physically assaulted by their purchasers, and a third of these experienced purchaser assaults at least several times a year. Twenty-three were beaten severely enough to have bones broken; and two were beaten into a coma.[15]

In another survey of 55 victims/survivors of prostitution in the state of Oregon, 78 per cent reported being raped by pimps and male buyers an average of 49 times a year. Eighty-four per cent were the victims of aggravated assault, often requiring

14. Sachi Sri Kanta, *Prostitutes in Medical Literature*, Westport, (Conn., United States), Greenwood Press, 1991.
15. Ruth Parriott, *Health Experiences of Twin Cities Women Used in Prostitution: Survey Findings and Recommendations*, 1994. Available from Breaking Free, 1821 University Avenue, Suite 312 South, St Paul, Minnesota 55104 (United States).

hospital emergency-room treatment; 53 per cent were sexually abused and tortured; and 27 per cent were mutilated.[16]

The dawning recognition that the etiology of HIV/AIDS in women is sexual intercourse with infected men has begun to realign the dominant view of prostituted women as vectors of sexually transmitted disease. The twelfth AIDS conference, held in Geneva in 1998, and World AIDS Day in December 1998, presented recent data on the prevalence rates and geographical and gender-specific spread of the disease revealing that 'men drive the AIDS epidemic'. Men who use infected women in prostitution then infect their wives, sexual partners and, often, their sexual prey. Women and girls are, consequently, contracting HIV primarily from men and at a faster rate than men and boys.[17] Women and girls in prostitution are more frequently exposed to infected men and thus at greatest risk.

In numerous countries, women and girls in prostitution have the highest rates of HIV in the population. In Cambodia, almost one in two prostituted females test positive for HIV.[18] More than one quarter (26.5 per cent) of urban prostitutes in Burma (Myanmar) have HIV, a rate many think is much higher.[19] Fifty percent of prostituted women in Bombay's red-light district are infected with HIV/AIDS.[20] In 1992, approximately 25 per cent of prostituted women in Thailand were HIV-positive, yet more than

16. Susan Kay Hunter, 'Prostitution is Cruelty and Abuse to Women and Children', *Feminist Broadcast Quarterly*, Spring, 1993.
17. Martin Foreman, 'A Global Epidemic Driven by Men Puts Women at Risk', *The Gazette*, Montreal, 28 November 1998, p. Bl.
18. 'Prevention Cambodia Launches Offensive in War Against HIV', *AIDS Weekly Plus*, 2 November 1998, p. 2.
19. Cesar Chelala, 'Burma: a Country's Health in Crisis', *The Lancet*, 15 August 1998, p. 556.
20. Robert I. Freidman, 'India's Shame: Sexual Slavery and Political Corruption Are Leading to an AIDS Catastrophe', *The Nation*, 8 April 1996.

half of single, urban, Thai men and one-third of married urban men had used prostitutes that year.[21]

Most women in prostitution are required to service multiple customers per day. Studies have documented that some women have even had to service as many as 100 men per day, orally, vaginally, and anally.[22] Sexually transmitted diseases of the lower and upper reproductive tract including syphilis, genital herpes, chancroid, trichomoniasis, chlamydia, and gonorrhea further increase the HIV transmission rate in women two to tenfold,[23] and are in themselves enormous health burdens.

HIV/AIDS is both a stark disease burden and a biomarker of the sexual condition of women and of male sexual consumption. Our hypothesis is that the highest rates of HIV/AIDS in the world today exist in centres of sex tourism, in the military occupations, and in societies and subcultures that accept and/or condone male sexual exploitation and female sexual subordination. When these landscapes of sex trafficking and prostitution are further riven by economic collapse, as in South-East Asia and the newly independent states of the former Soviet Union, we see the emergence of new and the re-emergence of 'old' sexually transmitted diseases.[24]

---

21. Russell W. Belk et al., 'Sexual Consumption in the Time of AIDS', *Journal of Public Policy and Marketing*, Vol. 12, No. 2, 22 September 1998, p. 197.
22. Kanta, op. cit., p. 173.
23. Jodi Jacobson, 'Women's Reproductive Health: The Silent Emergency', *Worldwatch Paper 102*, p. 30, Washington, D.C., Worldwatch Institute, 1991.
24. United Nations Population Fund, *Southeast Asian Populations in Crisis*, New York, UNFPA, December, 1998. See also Murray Feshbach, 'Dead Souls', *The Atlantic Monthly*, January 1999, pp. 26–7.

Conclusion

As with other forms of violence against women, eradicating the health burden of prostitution and sex trafficking entails addressing but also going beyond its health effects. To address the health consequence of sexual exploitation, the international health and human rights community must understand that prostitution harms women, and that in addition to needing health services, women must be provided with alternatives to prostitution. A fuller health response also requires that the male purchaser and the sex industry – the true vectors of sexual disease and sexual abuse injury – be prohibited from promoting, profiting from, or engaging in the sexual exploitation of women and children; and that ideologies which promote sexual exploitation and predation be publicly challenged and confronted.

Women's health in mid- and later life

Margaret Baltes* and Elisabeth Steinhagen-Thiessen**

With the increase in life-expectancy and the growing number of older women, their health concerns require particular attention.
(Fourth World Conference on Women, Beijing, 1995)

Health concerns are central to one's well-being at all stages of life, but it might be suggested that, in general, they become more acute as one grows older. For women, there are certain common experiences in mid-life and old age that can influence general health: menopause; 'empty-nest syndrome'; widowhood or death of a partner; and caring for the younger generation. Furthermore, as they age, women, like men, are confronted with the growing risks of chronic illness. Psychological, biological and life-style factors all play a role in determining the quality of health, though in differing degrees and at different times.

* Research Unit for Psychological Gerontology, Benjamin Franklin Medical School, Freie Universität Berlin (Germany).

** Director, Lipidambulance and HEIP-Apherese, University Clinic Charite, Medical Faculty of the Humboldt University, Berlin (Germany).

Researchers have recently reminded us that health is not merely a medical issue, but an issue central to the possibility of living a good life (Baltes and Horgas, 1998). From a feminist perspective, such considerations can entail ethical questions about the very definition of states of health and illness. For example, menopause can cause such vast shifts in all aspects of a woman's life that it could be argued that a more holistic definition than the one many medical professionals would now accept is called for. Further ethical questions raised from the feminist perspective concern equal access to health care, especially in cases where considerations must be made about screening procedures for conditions such as osteoporosis, breast cancer and ovarian cancer, or in cases where the accessibility of high-cost treatment to women of low socio-economic status is in question.

Menopause

Usually around the age of 50, biological and psychological changes in women are determined by a gradual decrease in the function of the ovaries (Bellantoni and Blackman, 1995). The gradual hormonal change results mainly from a decrease in levels of oestrogen, which influence important metabolic pathways, such as lipoprotein and bone-mineral metabolism. The general physical transition experienced during menopause may be accompanied by many conditions: vasomotor instability; atrophy of the epithelia, the urogenital tract or skin; a decrease in breast size and in bone density, with this latter leading to osteoporosis; and adverse changes to lipids and the cardiovascular system, with this latter condition increasing the risk of cardiac disease. These metabolic changes have enormous consequences in terms of morbidity and the degree of disability in old age. Low levels of oestrogen influence atherosclerotic development, resulting in numerous diseases typical of old age: stroke, myocardial infarct, peripheral artery disease, etc.

The extent of these metabolic changes in women varies considerably and can be enhanced or minimized by psycho-social factors (Avis and McKinlay, 1991). Menopause marks a transition in a woman's life, and a transition that is not only physical. It may be accompanied by stress and health problems, and it may entail a reappraisal of goals, self-image and relationships. There are also great cultural variations in the experience of menopause as a health and/or psychological problem.

The stress introduced by menopause may be exacerbated by other factors, such as the demands for care from both the older and the younger generation (Borchers, 1997). Where adult mortality remains comparatively high, and especially in those countries dramatically affected by the HIV/AIDS epidemic, middle-aged women may be called upon to support grandchildren who have been orphaned. In one area of Uganda almost 15 per cent of all children aged 14 years or younger had lost one or both parents, and approximately half of these losses were associated with HIV infection (Nalugoda et al., 1997).

Increasing longevity, in developed and developing countries alike, means that a growing number of middle-aged women have to care for elderly parents or other relatives while still, perhaps, looking after adolescent children or the babies of married sons or daughters. The burden may be particularly acute for women when husbands or adult children migrate to work in cities or even in different countries, leaving them to handle not only family care and housework but, in many instances, new tasks on the family farm. In Yemen, for example, women have been reported participating in activities such as ploughing that used to be traditionally reserved for men (Makinwa-Adebusoye, 1993).

Chronic illness

From mid-life and increasingly into old age, diseases due to lifestyle, biological and genetic factors become manifest and affect autonomy. Diseases involving bone metabolism often lead to disabilities that reduce the quality of life because of a lack of mobility and a need for care, and may result in social isolation. Muscle and bone tissue are closely related, and both are involved in the health problems of the locomotor system in older women. Remaining active is itself preventive health care. Life-style is also significant, as typical risk factors for osteoporosis include – in addition to genetic disposition and loss of oestrogen – low calcium intake throughout life, low physical activity and heavy smoking. Bone loss, especially in women, begins with the end of the third decade.

The majority of the diseases of the cardio-vascular system are due to the development of atherosclerosis, which is also latent over decades. Contributing factors, of which there are many, include high blood pressure, lipoproteins, diabetes mellitus, cigarette smoking, genetic disorders, obesity and lack of exercise.

Impact on health

Physical health

In terms of longevity, morbidity and disability there are significant gender differences. The higher life expectancy of women leads to the feminization of old age, and differentials in disease to greater disability among older women.

In a recent study known as the Berlin Ageing Study (Baltes and Mayer (eds.), 1999), the average number of diseases found in participants aged from 70 to 103 years was five. These were mostly chronic and negatively-interacting diseases, with intermittent acute phases leading to increasing deterioration. Diseases of the locomotor system are the predominant illness of elderly women. The Berlin study showed that ageing was accompanied by a loss in the strength of hand grip, a loss of mass in the back muscles and a decrease in the distance a participant could walk. At the age of approximately 70 years, a woman's continuous loss of bone mass intensifies. The strength of hand grip is essential for the activities of daily living; loss of back and chest muscle mass can affect standing and walking. Osteoporosis can generally result in fractures; in the elderly, and typically in elderly women, vertebral fractures are common, but hip fractures are often the most serious. Other impairments associated with ageing – such as loss of vision or co-ordination, deficits in the central nervous system and cardiovascular disturbances – also contribute to the high incidence of hip fractures among the old, although osteoporosis remains the predominant factor.

Slightly fewer women than men suffer ischemic heart failure, and women suffer from coronary heart disease and myocardial infarction, on an average, ten years later in life than men. In addition to minimizing cardio-vascular risks across society, exact guidelines for hormone replacement therapy (HRT) for women in menopause are needed. For women, HRT is an important issue in mid-life and in the later years, but it is still controversial even in Western countries. Current clinical trials, such as the Women's Health Initiative of the United States National Institute of Health, will establish the benefits and risks of hormone therapy and other interventions.

Functional health

Physical disabilities and the need for assistance in everyday tasks are major concerns in later life. When the need for assistance was measured in the Berlin Ageing Study, women aged 70–84 years needed significantly more help with such things as shopping and transport than men. Among women aged 85 and over, the need for help was even greater. Nevertheless, 66 per cent of the entire population above the age of 70 claimed to be completely independent. One-third of the men and almost half of the women in the study mentioned problems with reading or watching television. The study also tested balance, co-ordination and gait.

Twice as many women as men were found to have complete disability. These results are similar to those of other studies (Ory and Warner, 1990), and demonstrate the same phenomenon: women live longer than men, but have a significantly higher degree of disability and a higher rate of morbidity.

Mental health

Clinical depression is less frequent in old age compared to earlier in life. Data from the Berlin Ageing Study indicated a 24 per cent prevalence rate for psychopathology among the population aged 70 and above. However, general symptoms of depressive behaviour show an increase after middle age. In general, women are diagnosed with higher rates of depression than men, in a ratio of about 2 : 1 (Culbertson, 1997), but these findings are equivocal depending upon ethnicity, culture, socio-economic background and other characteristics of the group studied (Kessler, McGonagle and Zhao, 1994; McGrath et al., 1990; Nolen-Hoksema, 1990; Klerman and Weissman, 1989).

Dementia – of which the most frequent type is Alzheimer's disease – is the most common mental illness of old age; its general prevalence in society is significantly related to age. Gender differences in dementia have been frequently reported, but the direction of the effect is not always clear. Some reports (Gatz et al., 1995) indicate that cognitive impairment among those aged 65–74 years is twice as common in men than in women; however, at 85 years and older, women have been shown to have higher rates of impairment (Helmchen et al., 1998). These differences might be due to the type of dementia assessed; Alzheimer's disease is more frequent among women, while vascular dementia is more frequent among men.

Conclusion

Generally speaking, adults have at their disposal a range of resources that help offset the detrimental results of a decline in health, at least until extreme old age, when there seems to be a trend towards lower life-satisfaction. The effects of losses experienced by women in mid-life and old age are very specific. Some may be related directly to biological sex differences, such as the greater incidence of osteoporosis in women. Furthermore, as soon as we consider general issues related to gender, we discover that psycho-social conditions are different for men and women and lead to distinct effects in each. Gender differences are intricately entwined with major socio-economic factors, even in later life. As long as we do not make major advances in changing the social conditions of women – by removing social barriers – we will not address the causes of many health issues. In this regard, the most pressing ethical issue is the continuing discrimination against women.

Case study

The impact of widowhood on older women's health and well-being

Susan Feldman and Rosie Beaumont*

The enjoyment of good health and quality of life is the inalienable right of all women throughout their life cycle. Yet, there is a fundamental and disturbing link between a woman's health and well-being and her ability to maintain a visible social, economic and

* Alma Unit on Women and Ageing, Centre for the Study of Health and Society, University of Melbourne (Australia).

political status within her community, specifically as she ages and after the death of a spouse. In particular, the ageing process and widowhood have serious implications for many women far beyond the marking of a new stage in their life cycle. These life changes may signal significantly decreased physical and mental health coupled with inequitable health-resource allocation, financial insecurity and social isolation. In both developed and developing nations, entrenched cultural traditions and attitudes may marginalize and discriminate against women who are both economically and socially dependant or perceived to be inferior. In order to understand and address the health needs and rights of older widowed women from a global perspective, emphasis must be placed on the protection of their human and legal rights, their social and economic status and the changing nature of the family unit.

Widowhood

While the ageing of the world's population is recognized as a key global issue, that such an issue is a female concern is a point that is less commonly made. At the turn of the century, the average life expectancy for both women and men in developed countries was less than sixty years. By 1994, life expectancy for females had not only extended to beyond 80 years but had exceeded the average male life expectancy by approximately seven years (United Nations et al., 1991). International studies indicate that, if this trend continues, by 2001 women over 80 years will outnumber men by more than two to one. In developing countries these trends are similar, with female life expectancy ranging from 50 years into the mid-60s in the least-developed countries, and into the 70s in countries undergoing rapid development (Bonita, 1996). Indeed, it has often not been recognized that the populations of developing nations are ageing much more rapidly than those of the developed world (United Nations et al., 1991).

As a result, a significant proportion of the ageing female population – up to 70 per cent in some developing nations – is comprised of widows (Owen, 1996). For many of these older women, widowhood can lead to a serious and immediate deterioration in the quality of life due to ostracism, disfranchisement and even poverty (Lopata, 1987*b*; Owen, 1996). For others, however, some recent research suggests that widowhood can be a period when women find a renewed opportunity to develop contacts within their communities (Lopata, 1987*a*; George, 1980).

Negative images and cultural discrimination

The negative images and poor attitudes towards older women in general, and widows in particular, are international and cross-cultural phenomena (Owen, 1996). Widows from developed countries attest to the effects that negative stereotypes of ageing and singleness have on their lives. Older women report that they often suffer from a lack of acknowledgement, from patronizing or dismissive treatment and from the subtle impact of social stigmatization (Lopata et al., 1982; Martin-Matthews, 1991; Markson, 1994). Older widows report that they often become excluded from social and cultural activities, and that this leads to a more general isolation. Although the degree of such ostracism varies significantly, the general experience is nevertheless echoed by women from a wide range of cultural backgrounds in both developed and developing countries. Such an experience can have profoundly negative effects on an older widow's loss of self-esteem and her relationship with family and community, leading to an increase in stress-related health problems (van den Hoonard, 1997).

The vilification of older women in some cultures appears to be linked to beliefs about the motivations and worth, both social and economic, of older women and to confusion about the social role they should play. Widows also present challenges to the traditional female roles of mother and wife and are often seen as significant economic threats to other younger and married women (Keith, 1989; Owen, 1996; Russell, 1987).

While many developing countries constitutionally guarantee the equality of men and women, cultural traditions that may humiliate and ostracize women often override official rights (see pp. 97–106). The general reluctance of official regulators or justice systems in some countries to 'interfere' in private spheres and uphold women's human and political rights (see pp. 107–116) has meant that discrimination against older women and older widows can be significant.

One of the most pressing problems for older widows may be their relationships to their immediate and extended families. Their experiences often run counter to the accepted view that the family will always provide security and support to its older members, particularly in 'traditional' societies where older people are revered (Desjarlais et al., 1995). One example of this type of situation is Nigerian customary marriage law, which states that 'the beauty of a woman is her husband' (Bonita, 1996). When a husband dies his widow is seen as impure and unclean and she must endure traditional mourning practices, which can be harmful to her health. Many African and some Indian castes practise a custom similar to 'levirate', by which a deceased man's brother or other close male relative must assume the care of his surviving wife, children and land. While this practice can ensure ongoing economic and social support for a widow, it can also lead to violence and homelessness (Owen, 1996). The patrilineal line is so strong that widows who do not have male children are particularly susceptible to outright rejection by their deceased husband's family. The culturally accepted but legally illegitimate practice commonly known as 'land grabbing' is also a problem experienced by widows in virtually all developing countries. 'Land grabbing' is said to occur when a deceased man's relatives physically and emotionally abuse his widow and proceed to evict her forcibly from the home and land acquired by her husband, despite her legal rights to hold such property (Owen, 1996; Lopata, 1996).

Abuse of the elderly

Although there appears to be a stark difference between the impact of cultural vilification of older widows in developed and developing countries, recognition must be given to the growing problem of abuse of the elderly in developed nations. The issue first arose in the United States in the 1970s. The issue is now being confronted in Europe and Australia, with most research having been undertaken in the last five years (Office of Ageing, 1994). Older women, particularly widows, are vulnerable to abuse and violence, an occurrence that is under-reported in developed countries and often ignored by the broader community (Office of Ageing, 1994). The abuse may be psychological, financial and even physical. Such abuse of older widows may occur in a range of situations involving family members, neighbours, friends or even those entrusted to provide professional care.

Poverty

The Beijing Women's Conference (1995) reconfirmed that 'for the majority of women, particularly those who face additional barriers, continuing obstacles have hindered [an] ability to achieve economic autonomy and to ensure sustainable livelihoods for themselves' (United Nations, 1995*a*). This statement is of particular relevance to older women and widows throughout the world. Recent studies concerning the required needs to ensure health and maintain the quality of life of older American and Australian widows have reported that these women experience particular difficulties managing their finances and living on low incomes. Only one-third of the widows involved in an American study had held bank accounts in their own names during their married lives, while less than half had planned for their financial survival in the event of their being widowed (O'Bryant and Morgan, 1989).

There is also a need to achieve a global and systematic recognition of older women's contributions to production in the domestic market, informal economic activities and community life (Hoskins, 1992). Older women perform a significant proportion of the world's unpaid labour (Bonita, 1996). Indeed, the global exploitation of women in general as unpaid care-givers and domestic workers cannot be over emphasized (Bonita, 1996). This burden of exploitation may be further exacerbated by the ageing process or by widowhood. As economic pressures and patterns begin to alter the social fabric of virtually all communities, ageing women are increasingly taking up the responsibility of caring for ageing spouses, young children and sick or disabled adults.

In countries where social-security and pension schemes exist, they are overwhelmingly predicated on an individual's participation in continuous remunerated employment. Because the pattern of women's working lives can be interrupted by pregnancy and family responsibilities, they often do not meet the criteria for eligibility for benefit schemes. Furthermore, high levels of illiteracy or a lack of experience and training in financial management can mean that many older women and widows do not receive the vital financial support to which they are entitled (Global Link, 1995). One example of this situation are the many Indian widows who, although eligible for state pensions, lack the literacy skills to complete a complex bureaucratic application (Owen, 1996).

Access to health services

Many older women have never experienced financial independence and, after the death of their husbands, it is not uncommon for women to experience a dramatic decrease in personal income or even an inheritance of debt. The consequent financial insecurity has a direct impact on these women's ability to access appropriate health care. Preliminary analysis of data from a current longitudinal study of Australian women's health has revealed a correlation between widowhood and a drop in private health insurance, with significant numbers of older Australian widows indicating private health insurance as an item they cannot afford. The consequences for these women include absolute reliance on public health care and potentially longer waits for services (Byles et al., 1999).

For widows in developing countries, difficulties in accessing appropriate and affordable health services are caused as much by a chronic lack of services, particularly in rural regions, as by financial restrictions. Furthermore, a lifetime of poor nutrition, physical labour and numerous pregnancies leave many women with significant health problems in their old age:

> In rural areas of developing countries less than two-thirds of the population have access to safe water; less than half have access to sanitation. . . . Many aging women . . . have been left in small villages when younger people move to urban areas. . . . The great majority of aging women lack the means to pay for even basic health care. Using part of their limited income to meet the cost of health care would exacerbate the deprivations that contribute to their poor health (Bonita, 1996).

Internal migration

The economic pressures on developing countries, particularly those undergoing rapid industrialization, have led to rapid urbanization. This is fuelled by the arrival of young people from rural areas seeking better employment opportunities. Such a shift in lifestyles often results in older widows being left behind by their children, who may very well be the only ones providing care or financial security. In cases such as these, older widows may become reliant on the goodwill of neighbours or other relatives to maintain their lifestyle and land. Alternatively, they are forced to leave their homes, give up their status as owners of land and livestock and follow their children to the cities, where they often live in ghettos or slums (World Federation of Mental Health, 1993).

Care and support

In both developed and developing regions, the basic family structure is undergoing changes that diminish the ability of its members to provide security for the elderly. The increased participation of younger women in the workforce, urbanization and enforced or voluntary reductions in birth rates all take their toll. In Asia, for example, it could be argued that a crisis is looming due to a possible future dearth in the number of traditional care-givers: the Chinese government's 'one child policy' has dramatically reduced the ratio of young family members to elderly people in need of support. The dramatic demographic shift created by this policy, referred to as 'the 4-2-1 syndrome' – four grandparents, two parents and one child – is placing increasing pressure on single children to care for several elderly family members (World Federation of Mental Health, 1993).

Increasingly, children are choosing not to pursue the path of care-giver and consequently older women, particularly widows, are being left with few if any alternative means of support. This lack of support and care for older widows is often compounded by the continuing burden of care that they may face as grandparents and, in some instances, as the only surviving adult member of a family. As one critic remarks: 'The continuing burden of AIDS may well fall most heavily upon the grandmothers in certain developing countries. . . . In a family situation where both parents and some of the children are dying of AIDS, the surviving responsible person is likely to be the grandmother, who is likely to be the most active, fit and competent person to manage the family affair' (Bonita, 1996). More often than not the grandmother will be a widow.

Conclusion

In the past the health needs of women have been recognized largely in relation to their capacity to reproduce (Dennerstein et al., 1993). However, there are general issues associated with both ageing and widowhood – poverty, isolation, exploitation and the inequitable allocation of resources – that must be addressed. We must recognize the impact these issues have on the ability of older widows to enjoy good health and a decent quality of life. It is important that the health and well-being of these women be protected by society so that current widows and married women who experience widowhood in the future may maintain their social, political and economic rights.

Women and occupational health

Penny Kane*

Formulate special policies, design programmes and enact the legislation necessary to alleviate and eliminate environmental and occupational health hazards associated with work in the home, in the workplace and elsewhere.
(Fourth World Conference on Women, Beijing 1995)

Perhaps no area of women's health has been so neglected as that of health issues related to women's work. There are a number of reasons for this neglect, of which the most fundamental has been the difficulty of defining women's occupations within the existing frameworks used for collecting data. Women's occupations – more so than those of men – are fluid, multi-dimensional and frequently invisible. As a result, a recent report on women's occupational health could do little more than illustrate the dimensions of the problem (Kane, 1997).

* Consultant on health, population and family planning; University of Melbourne (Australia).

Defining women's work

Much of women's work goes unrecognized, uncounted and unpaid: work in the home, in agriculture and in the marketing of home-made products, for example. Women in sub-Saharan Africa provide over 70 per cent of all agricultural labour (United Nations, 1996*a*); in South Asia they provide more than half of such labour (United Nations, 1995*e*). For the Baluchistan women of Pakistan, the tasks associated with 'housework' include fetching water from as far as ten kilometres two to four times a day, collecting wood and fuel for heating and cooking, managing home and children and supplementing family income by wool-sorting, embroidery and straw-weaving, among other tasks. Yet the official labour force participation rate for Baluchi women, as reflected in the 1981 census, was 1 per cent. Figures such as these reflect substantial injustice in the assessment of women's contributions to these societies.

The work of women in industrialized countries is also inadequately recorded by most conventional employment statistics. In Australia – one of the first countries to undertake detailed time-use studies and incorporate the findings in national accounts – women over the age of 15 were found to spend twice as much time as men on unpaid household activities. That unpaid housework, together with voluntary and community work, was equivalent to 38 per cent of GDP; in other words, women's unpaid activities could be assessed as increasing the country's domestic product by almost two-fifths.

Women may undertake paid employment at home, or combine part- or full-time paid employment with household work and the care of children, the sick or the elderly. Different countries define part-time work, casual work and unemployment in different ways, and this further limits opportunities for international comparisons. In addition, women move in and out of the workforce during different life stages; within the paid labour force they may have a variety of different occupations in succession.

Tracing connections between their various 'professional' activities and morbidity and mortality is thus extremely difficult. Even within the paid workforce, women are disproportionately concentrated in the informal sector, beyond the scope of industrial regulations, trade unions, insurance or even data collection. Thus, health risks to women workers are less likely to receive recognition, and women are less likely to be protected from potential occupational hazards or to receive compensation for health impairment suffered through their work.

Occupational mortality and morbidity data

The limitations of mortality as an indicator of occupational health risks, for men and for women, are well-known. Considerable time may pass between exposure to the risk and any negative effect on health, or, ultimately, death. Early manifestations of ill health may lead to a change of occupation. For women, whose activities in the paid workforce are irregular, the difficulty of identifying a causal relationship between occupation and fatal health risks is compounded.

Accidents may seem to present the simplest relationship of cause to effect; but, in fact, even this is doubtful. For example, Russian data concerning mortality provide information on accidental deaths that is divided into two categories: 'work-related' and 'not work-related' (Mesle et al., 1996). Women's mortality rates are concentrated within the category 'not work-related', but we have no way of knowing how many of the accidental deaths documented in this category resulted from activities such as cooking, cleaning windows or taking a child to school.

In Russia (Mesle et al., 1996), as is typical of all societies (Kane, 1991), far more men than women suffer from accidents of all types. However, detailed data concerning disability, where available, show that within this overall pattern, women may nevertheless be particularly vulnerable to specific occupational accidents. In Malawi, for instance, more women than men have lost toes from one foot or been scalded seriously enough to be classified as disabled, and almost as many women as men have been badly burned (Malawi National Statistics Office, 1991). Digging and hoeing the fields, as well as cooking, are clearly high-risk occupations for women in that country, and perhaps elsewhere.

Thus, data concerning morbidity at least offer clues to potential occupational health hazards, even if they can seldom document the full extent of any problem attributable to a single cause. Longitudinal studies and record linkages would be required for a better understanding of the many occupational health risks women experience, especially those that are potentially disabling and even fatal.

However, such possibilities are generally limited to industrialized countries. In developing countries, even statistics documenting the cause of death are either unavailable or based upon samples that may be far from representative, and information on morbidity is virtually non-existent. The Malawi example mentioned above is a rare exception. Consequently, there is really no epidemiology of occupational health; we simply do not know the incidence or the prevalence of the occupational health problems of either men or women.

High-risk occupations

There are certain types of occupations which, throughout the world, are predominantly undertaken by women, and each has its particular hazards. Thus, while not all women undertake paid employment, few can escape household labour. What such labour involves varies with income, class and culture; but because the home is such a basic feature of everybody's lives, the routine risks encountered while working in the home are easily overlooked.

The risks can be found, nevertheless, in the chemicals used for cleaning or fuels used for cooking, as well as in the terminology of some medical conditions, such as 'housemaid's knee'. They can also be found in the isolation of many wives, especially those who have married into an unknown family, perhaps far from their original homes, or those whose husbands leave home to find work in the city, or even in another country. And they can be found in the domestic violence – perpetrated by husbands or other family members – which many women face as they go about their daily tasks.

Those women whose husbands work away from home are also at a greater risk of catching diseases such as tuberculosis and especially sexually transmitted diseases (Orubuloye, Caldwell and Caldwell, 1994). When women themselves migrate, it is often to badly-paid work in exploitative conditions; international migration may bring problems of isolation and stress as well as difficulties in accessing health information in a new language.

Household labour also frequently involves caring for other family members: children, the sick and the elderly. Even where men share this work, it is usually the woman who is the primary carer, and she may suffer additional health risks. These may be physical – for instance, muscular problems caused by lifting – or involve extreme tiredness, stress or depression.

Women predominate, too, as paid providers of care – for instance, as health workers of different types – and encounter many of the same problems in this role. One-third of people providing 'home help' services and personal-care workers who responded to an British trade union survey had received no training of any kind and fewer than half had been trained to lift people safely, even though their tasks involving high risks of back injury included bathing people, lifting them on and off commodes or in and out of bed, etc. (NUPE, 1993).

Millions of women today, as in the past, find work in what is called the 'sex trade' (Bloor, 1995; Bolvary and Vacziz, 1996; Pan, 1996); often it is the only means of employment available to them (Adnan, 1993; Encloe, 1995). While the occupational health of women in the sex trade varies with the meanings, customs and context of sex work, the degree of control they can exercise over their lives is the crucial determinant of their health status. Risks – especially of violence and disease – are incurred not only through customers, but also through those who manage the circumstances of the trade: brothel owners, pimps and the police (Larsen, 1996; Lowman and Fraser, 1995). Repressive legislation may serve only to drive the women away from health agencies and health interventions.

The effects on health of women's multiple roles are still poorly understood. The impact of different roles and responsibilities at different stages of the life cycle remain largely unexplored. If much of the current literature on women and paid work, especially that concerned with mental health, is ambiguous or contradictory, it frequently reflects inadequate research and an unjustifiable level of generalization about women's lives.

Women's activities and their health risks

Identifying occupational health risks involves understanding precisely how an activity is carried out and by whom. Cooking, for example, may imply very different types and levels of hazard depending upon a number of factors including the type of stove and fuel in use. Much of the reported high level of respiratory disease among women in many developing countries may be related to cooking practices. A study of rural kitchens in India found concentrations of pollutants 100 times the levels acceptable to the World Health Organization (World Bank, 1996). Domestic smoke combined with the inhalation of silica particles while hand-grinding maize between rocks in South Africa has been linked to a form of pneumoconeosis called Transkei silicosis (Grobbelaar and Bateman, 1991).

A simple occupational category is seldom sufficient to tell us who undertakes which activities. Agricultural workers, for example, may dig and hoe and apply fertilizers and pesticides, but not all the workers will perform all of those tasks. Among Thai villagers, one study found men and women equally likely to undertake insecticide spraying on cotton, but women less likely than men to spray vegetables (Sirisambhand and Gordon, 1987). More women than men applied fertilizer to both cotton and vegetables. In Canada, among professional cleaners, males mop and females dust (Courville *et al.*, 1996). Where the tasks are segregated by gender, the health implications of a particular occupation for men and women may be very different.

Gender and occupational health

Gender segregation is likely to vary between cultures. But, in general, there are very few activities that can universally be described solely as 'women's work'. As a result, occupational health risks are seldom universally confined to either men or women alone; nevertheless, they are only likely to be fully understood, and confronted, in the context of a gender-specific analysis of occupational health.

Such an analysis requires a recognition of the extent of gender variations and careful controls for biological and social characteristics that may affect health. In South Asia, for example, poor nutrition may be a more important factor in some types of occupational health risks than simply being female. And where anthropometric measurements related to wrist anatomy and physiology are taken into account, carpal tunnel syndrome is not, as is often assumed, related to gender.

The effects of potential occupational hazards on women's reproductive health have probably been the major focus of concern in the health of women workers. This concern has increased in recent years as more environmental hazards are identified and as more women enter the paid workforce. A large range of occupational reproductive hazards has been documented.

Strenuous physical work, stress and a poor physical environment, as well as the presence of a variety of dangerous chemicals, have been found to affect women's fertility. Solvents, organic pollutants and pesticides, as well as heavy workloads and postural factors, have been linked to spontaneous abortions and other foetal problems. A large number of possible risks still require further examination; more noticeable perhaps is how little research has been done on occupational hazards specifically affecting men's reproductive health.

Laws and policy in occupational health

Legislation to protect against the eventuality of problems during pregnancy has been one response to the occupational health concerns of women. However, the drive to create such legislation may well result in potential reproductive hazards for male workers being ignored. It may also result in legislation that is too restrictive, depriving women of an income on the assumption that they are all perpetually pregnant. A classic example is work with lead, from which many countries, including the United States, Canada and Australia, banned women for many years. In fact, toxicological responses to lead are similar in both men and women (International Labour Organisation, 1995).

A number of countries have had to re-examine their laws and regulations on occupational health because of new legislation on gender discrimination. And sometimes this process has led to unexpected additional benefits. In Germany, for example, bans on women working at night were found to be unconstitutional (Hoffman and Meyer, 1992). Among other restrictions affected by this ruling was the exclusion of women from the building industry. This in turn triggered examination of the high number of accidents and industrial diseases across the industry and a review of existing health and safety regulations.

In other countries, however, legislation cannot be enforced because of economic pressures. In 1988 the Vietnamese government listed heavy or harmful jobs forbidden to women, but, as has also been the case elsewhere, necessity along with the introduction of competitive economic practices even within state industries have meant that women continue to undertake the work (Morrow, 1995).

If other policies such as those establishing the criteria for participation in training programmes exclude women from learning to minimize the hazards they in fact face, women may paradoxically be at a greater risk than if there had been no attempt to protect them. For example, Russian women are excluded from training programmes in some 600 of the 5,000 recognized workers' occupations because of potential health risks (WHO, 1994*b*). But inability to enforce the regulations means that women end up not only doing the work – for example, driving tractors to apply pesticides – but without the requisite training and levels of skills, and on lower pay.

The outlook

Women are increasingly moving into paid employment, and staying for longer periods. In addition, they are interrupting this work less to have children. They are not working for pin money, but for the livelihood of their families. It could be argued that the economic success of the 'Asian tigers' relied heavily upon the increasing feminization of the paid labour force: the rates of participation of females in the workforce had increased rapidly in the region, though the jobs available to them were low-skilled and more poorly paid than those offered to men (Lin, 1993).

Women are also increasingly employed in the informal sector and, in some countries, such as Zambia, Honduras and Jamaica, they make up more than half of all workers in this sector (UN, 1995e). The expansion of rural industries, together with the restructuring of economies, transforms traditionally female occupations, such as sewing, into new opportunities for paid employment, which in many countries become male employment. Women, however, may be left struggling to farm alone. The

Economic Commission for Africa states simply that 'rural women's work doubles when men migrate to the cities' (Ware, 1981). Often that work involves tasks for which the women have no training: Turkish women have been reported as driving tractors in the absence of their husbands, and in Yemen women have been reported doing work such as ploughing that has traditionally been reserved for men (Makinwa-Adebusoye, 1993). Farming is recognized across the world as involving fairly high levels of accident and injury. The hazards are likely to increase when farm tasks are left to those who have never practised them before.

Women's occupational health therefore involves a range of issues of increasing significance. Legislation to protect the health of working women, however valuable, has considerable limitations and provides only a partial strategy with which to address those issues. Its impact is largely in the traditional formal sector; even there, as indicated earlier, it has sometimes been inappropriate and actually led to reductions in the equity of employment opportunities. Where legislation cannot be enforced, it is at best irrelevant and at worst may be counter-productive, leaving the woman worker at greater risk than she would have been without a law designed to protect her. Legislation may even function as a substitute for action: countries which have a notoriously poor record concerning occupational health are very likely to provide exhaustive documentation of legislation on women and occupational health (Dennerstein and Kane, 1999).

Justice and equity require that opportunities for work not be denied to women as a result of arbitrary assumptions about their possible health risks or the traditional division of labour. Attempts to 'protect' women by excluding them from certain allegedly hazardous activities would be better turned to ensuring that the risks were reduced for all workers. Where the attempt to 'protect' women is based on fears of damage to their reproductive system, the principle of equity demands that the potential danger to the male reproductive system also be explored and, if necessary, that measures be taken to protect both sexes. Equity also demands that women be seen other than in a strictly potential reproductive capacity: in other words, that only very specific hazards, for example to the health of a foetus, should limit the participation of a woman in the workforce – and even then only during her pregnancy.

An ethical framework going beyond legislation has to take into account the variety of women's work and ensure gender equity by classifying 'occupation' by content, rather than by context. Laundry work, for example, is the same activity, with the same range of potential hazards, whether performed by a wife, an informal-sector laundry woman or a cleaning company. To ignore the health needs of the first two types of worker (who are also more likely to be women) is to ignore basic principles of distributive justice.

Case study

Women working with women

Elizabeth McGregor*

Issues of bioethics, women's knowledge systems of plants and intellectual property protection are visible in the front-line research of Bente Huntley, a Cree Indian living in Northern Saskatchewan, Canada, who dedicated her Masters research to the recording of oral teachings and knowledge of plants held by her people. In undertaking this research, Bente was guided by two significant women: one is from the recognized western scientific community, the other is an Elder Cree medicine woman from

* Senior Analyst for Gender, Science and Technology, Industry Canada (Canada).

the Lac La Ronge First Nation. Several unique hurdles had to be faced and solved in the process. A protocol for researchers of oral traditions was developed with the Cree Elders and is outlined in her presentation. Furthermore, for the first time ever, special recognition was granted by the University of Saskatchewan 'Ethics Committee' to a non-traditional medicine woman as an external thesis advisor.

Issues of intellectual property protection must inevitably be considered in research that makes community knowledge systems visible. The traditional system of patents, now over 125 years old, was not designed to address knowledge systems developed over time and by communities, and, in the case of the Cree story-tellers, not written down but passed through generations by oral tradition. Yet the stakes are high.

In the health field, 80 per cent of the world's population is at least partly dependent upon traditional medicine and medicinal plants to treat their ills (Shelton, 1993) and at least 7,000 medical compounds in the Western pharmacopoeia – from aspirin to birth control pills – are drawn from plants (Mshigenio, 1990). The estimated value of medicinal materials could range from $35 billion to $47 billion by the year 2000. . . . Because the development of medicinal plants relies heavily on the knowledge of indigenous peoples and rural societies, concerns about equitable benefit sharing and intellectual property inevitably arise ('People, Plants and Patents', The Crucible Group, International Development Research Centre).

Moreover, culturally speaking, the concept of ownership over plant forms and nature is considered inappropriate to some societies.

The United Nations Declaration of Human Rights set forth a vision of social equity and justice in which dignity and equality of all people are respected without regard to race, colour, sex, language, religion, political or other opinion, national or social origin, property, birth and other status. However, this declaration ignored cultural rights as valued aspects of human rights (Lou Heber, 1999).

A guidebook, *Indigenous and Local Community Knowledge Systems in Animal Health and Production Systems – Gender Perspectives'* provides a road-map to these cultural and Indigenous people's issues, global recommendations, leading researchers and NGOs, field studies and research methodologies and active networks. Co-ordinated by researcher Catherine Hill, it was developed in collaboration with the World Council of Indigenous Peoples, the IICA and the World Women's Veterinary Association (WWVA) for the Fourth World Conference on Women (Beijing, 1995).

In 1995, the United Nations Commission on Science and Technology (UN-CSTD) through its Gender Working Group, drew international attention to bioethics and indigenous knowledge systems. It endorsed a *Declaration of Intent* and seven Transformative Actions on Gender in Science and Technology. Among these seven were ethics and indigenous knowledge systems. The Commission recognized that women are often the holders of traditional knowledge and called for preservation of local knowledge systems with particular attention to their gendered nature.

The work of Bente Huntley does just that. Working directly with the Cree Elders and the storytellers, Bente has developed a 'teaching module for schools' which preserves the stories on tape as told by the owner of the story for future generations.

The following section is her first-person account of the finalization of her work.

Bioethics, plants and the Cree storytellers

Blankets. Tobacco. Jams. Birch bark bitings. Eagle feathers. Sweetgrass. Chairs in a circle. Slides, tape recorder and projector. Ready to go. This was it. The last part of my master's project: my oral defense. It had been a long struggle. This was the last hurdle. I really needed this sharing circle to feel closure. My project was finished and was ready to be validated by my external advisor:

Elder Sally Milne, a Cree Medicine woman from the Lac La Ronge First Nations in Northern Saskatchewan. No 'western' academic credentials followed her name; yet Elder Milne was one of the few women who possessed the knowledge and expertise to validate my project. Certainly none of the western academics from the university could verify if my project was authentic; therefore, the ethics committee opened their minds more to the possibility that a person does not need to have the western academic credentials to possess knowledge. It was a big step. I felt humbled to have worked under Sally the past few years and honoured to have her validate the final project. I was also fortunate to have worked under a university advisor who understood my need and desire to promote Traditional Environmental Knowledge (TEK):[1] the knowledge that Sally was concerned about losing. Sally also feels, as I do, that TEK must be transmitted the way it has been for generations and that the sacred knowledge of the medicine lodge must be kept sacred. But the common knowledge contained in the many stories must be transmitted to the next generation, either orally or in written format. Thus many elders have given their stories to be printed for use in a good and right way.

For centuries, Indigenous peoples the world over have relied on the oral stories and knowledge of their environment for survival. 'For the past century or more, traditional knowledge was condemned as "heathen" or "folklore", but Aboriginal peoples' traditional teachings and knowledge are, at long last, being recognized as valid and valued. The resilience of Aboriginal peoples, the world over, has kept alive their oral traditions and teachings. Traditions and knowledge have long been underground and dormant, but many people are now starting to remember the grandfathers' and grandmothers' stories' (Huntley, 1999, p. 30).

I continue on this journey, which actually began many years ago with the oral stories of my childhood, because I am con-

1. TEK is term used by Martha Johnson in her book *Lore: Capturing Traditional Environmental Knowledge* (1992).

cerned about saving the knowledge of my ancestors: 'Knowledge that the forests and this planet are alive' (Paulinho, in Burger, 1990, p. 32). My research project concerns the transmission of Indigenous traditional environmental knowledge.

I have always been interested in the knowledge and stories my ancestors possessed about plants. So in September of 1973, I enrolled in a two-year Renewable Resources Technology course in Saskatoon, Saskatchewan. One of the four main components of the course was forestry. Upon completion of the course I worked in the area of silviculture for the Department of Northern Saskatchewan Forestry Branch. It was at this time that I had the unique opportunity to travel extensively through the North. This enabled me to understand the diversity of Indigenous people who live in northern Saskatchewan and to discover some of the common threads and knowledge about plants that weave through all their stories. For example, I was amazed to find how many other Indigenous people, besides my grandfather, smoked a form of *kinnikinik* (wild plant, usually bearberry or red osier dogwood mixed with tobacco). I wish to continue to explore the stories and knowledge of my ancestors. My fear that Indigenous oral history might be lost if steps are not taken to promote and renew it encouraged me to question how the oral traditions might be revitalized.

What is the Indigenous traditional environmental knowledge of the Cree people of north-central Saskatchewan? How is it that knowledge was passed on and how is it relevant for the classrooms today? I answered these questions through research and development, specifically through interviews and field work. Since this type of project deals with people's stories and lives, proper cultural protocol was followed and adhered to; it is the process, not the event, that is important in Indigenous cultures.

Cultural protocol varies from culture to culture and community to community. From my experiences, however, the cultural protocol process for most Indigenous cultures is quite similar; only the specifics vary. At the initial contact between Elder Vicki

Wilson (participant) and myself as researcher, an offering of tobacco was given. This was followed by the purpose my visit. I then sat and listened. This process sounds more straightforward than it really is. A relationship must be developed between the Elder and the researcher before the specific purpose of the visit can be addressed. Trust and respect do not happen overnight. The Elders tell us that things happen in their own time; therefore, it is imperative that I sit, listen and reflect on the words of the Elders. The period of 'sit and listen' might last for months, depending on the Elders' perception of the readiness of the researcher to absorb the knowledge that is to be shared. The 'reflection' period might also last for months, depending on the amount of time invested and the ability of the researcher to understand the concepts. The Elder controls the project in terms of the nature of and rate to which they share knowledge with the researcher. The researcher is responsible for his or her own learning and must accept that learning is not governed by time. It is important to remember the Elders' teachings and remember that there is a time and place for everything. Additional gifts of tobacco, blankets or food might be given at each session. The sessions could last from a few minutes to hours, so it was necessary to be prepared for any time line. The protocols are reflective of the teachings as well.

Usually, acceptance of the offering implies consent. No written approval is required. Indeed, in most cases, to ask for approval could be considered as an insult. The Elders' oral approval and acceptance of the offering is validation. Once the offering is accepted, no other permission is required. This has been the way of my people, which they have preserved and passed down for generations.

In order to obtain a deeper understanding of another way of knowing and learning, it is necessary to go to the sources of this knowledge: medicine people and Elders. The stories must be recorded and reproduced orally. In keeping with oral traditions and Indigenous values, this resource module is presented orally.

Alexander Wolfe (1989), a Saulteux storyteller, informs us that 'information and instruction were transmitted to us orally, in story form, by our old people. . . . If we are to preserve the stories that contain our history, we must restore the art, practice and principles of oral storytelling' (p. xv). Oral recordings provide an introduction to Indigenous ways of knowing and explain how Indigenous people used, valued, and identified the plants in their environment. Through their stories, the recordings pass on their history, their knowledge, and their contributions. The traditional stories were (and continue to be) an integral part of Indigenous cultures.

The interview with Elder Vicki Wilson was at the location of her choice. Elder Wilson (Walking in the Sky Woman) is an Assiniboine Cree from the Whitebear First Nations. She has been recognized as an elder for over ten years. Although she was a product of the residential school system, Vicki is a wise traditional woman who also follows the Dakota/Nakota teachings. Elder Wilson has spent numerous years working in the area of education. She has worked for Wonska Cultural School, SUNTEP, Saskatchewan Indian Federated College and Pine Grove Correctional Center for Women. In addition, she was one of the founders for Joe Douquette Survival School. She understands and knows about the strong connections Indigenous people had and still have with the natural world. She also feels that it is important to pass on the knowledge gained from generations past. Elder Wilson loves to tell stories and each story contains so much wisdom and knowledge. Of course, I turned to her as a participant in my project. Elder Wilson shares her stories of the medicinal uses of plants for burns, headaches, diabetes and infections. Her stories also teach about food sources and survival.

Anonymity was waived, for it is imperative that the stories remain the property of the people who told them. There was no attempt to disguise voices, edit stories, or change data. Part of the project consists of the importance of the oral traditions and teachings passed down through the generations. The validity lies within the stories of the informants. The interview and recordings of the stories vary in length because of the nature of the topic. Elder Vicki Wilson was given a copy of the oral stories (edited version) along with the corresponding slides, to review for authenticity and approval. In addition she was also given a draft copy of the final stage of the teacher's guide (including the rationale), oral introduction and project proposal. Any changes were then made and given back for final approval.

Traditional Environmental Knowledge of the Cree People of North-Central Saskatchewan is intended for use as a resource module for science education. Cultural awareness and the validity of Indigenous knowledge is promoted. The resource kit has three parts. Part A provides an introduction and rationale on the importance of preserving the traditional knowledge of Indigenous peoples. It also contains an overview on how to use the guide, as well as generic sample lessons for each story. For each sample lesson plan, the corresponding story and slides are included. Also included is background information on the identification and uses of each particular plant. Part B contains the slides of six different plants: birch, bearberry, dry ground cranberry, chokecherry, mint, cattail. Part C contains the oral stories and teachings of Elder Vicki Wilson. There are five stories recorded; each story is separated by four drumbeats. The oral tape also contains the background information about the plants and their uses.

I feel that a resource module on plants from an Indigenous perspective is needed in all schools. It is hoped that the oral stories will create more positive attitudes towards Indigenous peoples and their contributions to society. Our elders tell us the most important teaching is respect: respect for Mother Earth and all her creations. This includes the plants. Children should be taught to respect the plants they study, not to destroy plants needlessly; to appreciate and give thanks for anything they take from nature, and always to walk softly on the earth, for the earth gives them life. When listening to the Elder's stories it is important to listen closely for the teachings. As in all Aboriginal stories, there are many layers and many lessons. I hope everyone who listens to the wise words of the old ones will gain the valuable and rich knowledge that exists within their words, as I have.

References

Contents

AIKENHEAD, G. 1995. *Toward a Cross-cultural Perspective on Western Students Learning Western Science: Border Crossings.* Paper presented at the National Association for Research in Science Teaching, St. Louis, Missouri.

——. 1996. Science education: border crossing into the subculture of science. *Studies in Science Education*, 27, p. 1–52.

——. 1996. *Towards a First Nations Cross-cultural Science and Technology Curriculum for Economic Development, Environmental Responsibility, and Cultural Survival.* Paper presented at the International Organization for Science and Technology Education, Edmonton, Canada.

AITKEN, L. P.; HALER, E. W. 1990. *Two Cultures Meet: Pathways for American Indians to Medicine.* Duluth, Minn., University of Minnesota.

BATTISTE, M. 1986. Micmac literacy and cognitive assimilation. *Indian Education in Canada: The Legacy.* Vancouver, UBC Press.

BLADES, D. 1994. *Repetition of a Dream Catcher.* Paper presented to the 1994 JCT conference on curriculum theory and classroom practice.

BORGERSON, L. 1993. *Storytelling in Play: Upisasik Theatre Revisited.* Unpublished thesis. University of Saskatchewan, Saskatoon, Canada.

BURGER, J. 1990. *The GAIA Atlas of First Peoples.* New York, Anchor Books.

CAJETE, G. 1994. *Look to the Mountain: An Ecology of Indigenous Education.* Durango, Colo., Kivaki Press.

CHOMSKY, N. 1993. *Year 501: The Conquest Continues.* Montreal, Black Rose Books.

COBERN, W. W.; AIKENHEAD, G. 1995. Cultural aspects of learning science. In: K. Tobin; B. Fraser (eds.), *International Handbook of Science Education*, pp. 1–29.

CORSIGLIA, J.; SNIVELY, G. 1995. Global lessons from the traditional science of long-resident peoples. In: G. Snively, A. MacKinnon (eds.), *Thinking Globally about Mathematics and Science Education*. Centre for the Study of Curriculum and Instruction, University of British Columbia.

DELORIA, V. Jr. 1991. *Indian Education in America*. American Indian Science and Engineering Society, Boulder, Colo. (See chapters entitled American Indian Metaphysics; Power and Place; Knowing and Understanding; Traditional Technology.)

ERMINE, W. 1995. Aboriginal epistemology. In: Barman Battiste (ed.), *First Nations Education in Canada: The Circle Unfolds*. UBC Press, pp. 101–12.

GALLAS, K. 1994. *The Languages of Learning*. New York, Teacher College Press.

HAMPTON, E. 1995. Towards a re-definition of American Indian education. In: Barman Battiste (ed.), *First Nations Education in Canada: The Circle Unfolds*. UBC Press, pp. 5–46.

HART, E. P. 1978. *Study for Saskatchewan Schools: Field Study Report*. Regina, Canada, Faculty of Education, University of Saskatchewan.

HUNTLEY, B. 1995. *Aboriginal Pedagogy: The Oral Tradition and its Relevance for the Classroom*. Unpublished paper.

——. 1998*a*. Plants and Medicines: An Aboriginal way of teaching. In: L. Stiffarm (ed.), *Aboriginal Pedagogy*. Saskatoon, Canada, University Extension Press.

——. 1998*b*. *Traditional Environmental Knowledge of the Cree People of North-central Saskatchewan*. Unpublished master's project, University of Saskatchewan, Saskatoon, Canada.

INDIAN AND METIS ADVISORY COMMITTEE. Report. 1992. Saskatoon, Canada.

INTER PRESS SERVICE. 1993. *Story Earth: Native Voices on the Environment*. San Francisco, Mercury House.

JOHNSON, M. 1992. *Lore: Capturing Traditional Environmental Knowledge*. Hay River, Dene Cultural Institute.

KATZ, R.; ST. DENIS, V. 1991. Teacher as Healer. *Journal of Indigenous Studies*, Vol. 2, No. 2, pp. 23–6.

MCFADDEN, S. 1994. *Native American Wisdom*. Rockport, Mass., Element Inc.

MCIVOR, M. 1995. Redefining science education for Aboriginal students. In: Battiste, Barman (ed.), *First Nations Education in Canada: The Circle Unfolds*. UBC Press. pp. 73–100.

MILNE, S. 1995. *Living in a Good Way: A Cree Perspective to Education.* La Ronge, Saskatchewan, Curriculum Resource Unit, Lac La Ronge Indian Band Education Branch.

OGAWA, M. 1986. Toward a new rational of science education in a non-western society. *Eur. J. Sci.,* Vol. 8, No. 2, pp. 113–9.

POMEROY, D. 1992. Science across cultures: building bridges between traditional western and Alaskan native sciences. *The History and Philosophy of Science in Science Education,* Vol. 11, pp. 257–67.

——. 1994. Science education and cultural diversity: mapping the field. (To be published in *Science Education.*)

SASKATCHEWAN EDUCATION. 1995. *Science: A Curriculum Guide for the Elementary Level.* Regina, Canada, Saskatchewan Education.

——. 1989. *Report to the Minister of Education: Summary of Findings.* Sask. Canada, Northern Education Task Force.

SJOBERG, S.; IMSEN, G. 1995. Bridging traditional science and western science in the multi cultural classroom. In: G. Snively; A. MacKinnon (eds.), *Thinking Globally about Mathematics and Science Education.* University of British Columbia, Centre for the Study of Curriculum and Instruction.

SUZUKI, D. 1994. *Time to Change.* Toronto, Canada, Stoddart Publishing Co. Ltd.

SUZUKI, D.; KNUDTSON, P. 1992. *Wisdom of the Elders.* New York, Bantam Books.

VONTABEL, R. 1989. Two ways of knowing. *Caribou News,* Vol. 9, No. 2.

WA'THIONGO, N. 1986. *Decolonizing the Mind: The Politics of Language in African Literature.* London, James Currey.

WOLFE, A. 1989. *Earth Elder Stories.* Saskatoon, Canada, Fifth House.

WOLFE, J.; BECHARD, C.; CIZEK, P.; COLE, D. 1992. *Indigenous and Western Knowledge and Resources Management System.* University of Guelph, University School of Rural Planning and Development.

WOLFSON, E. 1993. *From the Earth to Beyond the Sky.* Boston, Houghton Miffin Company.

Methodology

ALASUUTARI, P. 1995. *Researching Culture: Qualitative Method and Cultural Studies.* Thousand Oaks, CA, Sage Publications, Inc.

ARCHIBALD, J.-A. 1993. Researching with mutual respect. *Canadian Journal of Native Education,* Vol. 20, No. 2, pp. 190–1.

HAIG-BROWN, C. 1992. Choosing border work. *Canadian Journal of Native Education,* Vol. 19, No. 1, pp. 96–116.

HOWELL, W. K.; FOX, S. L.; MOREHEAD, M. K. 1993. *Curriculum-based Evaluation*. Pacific Cove, California: Brooks/Cole Publishing Company.

KIRKBY, S.; MCKENNA, K. 1989. *Experience Research: Social Change Methods from the Margins*. Toronto, Canada, Garamond Press.

MILBURN, G.; GOODSON, I. F.; CLARK, R. J. (eds.). 1989. *Re-interpreting Curriculum Research: Images and Arguments*. London, Canada, The Falmer Press.

SANDERSON, J. 1991. Aboriginal pedagogy: an adult education paradigm. Unpublished master's project, University of Saskatchewan.

SEARS, J. T.; MARSHALL, D. J. (eds.). 1990. *Teaching and Thinking about Curriculum*. New York, Teacher's College Press.

SPRADLEY, J. P.; MCCURDY, D. 1972. *The Cultural Experience: Ethnography in Complex Societies*. Illinois, Waveland Press Inc.

STAKE, R. E. 1995. *The Art of Case Study Research*. California, Sage Publications.

ZAHARIA, F. 1996. *Kitomahkitapiiminnooniksi: Stories from our Elders*. Vol. 1, 2, 3, and Teacher's Guide. Edmonton, Canada, DHP Publishing Inc.

Plant and story resources

ASSINIWI, B. 1972. *Survival in the Bush*. Toronto, Canada, Copp Clark Publishing Co.

BROWN-ERICHSEN, C. 1979. *Medicinal and Other Uses of North American Plants.* New York, Dover Publications, Inc.

CADUTO, M.; BRUCHAC, J. 1991. *Keepers of the Animals*. Saskatoon, Canada, Fifth House Publishers.

——. 1991. *Keepers of the Earth*. Saskatoon, Canada, Fifth House Publishers.

——. J. 1994*a*. *Keepers of Life*. Saskatoon, Canada, Fifth House Publishers.

——. 1994*b*. *Keepers of the Night*. Saskatoon, Canada, Fifth House Publishers.

CARMICHAEL, L. T. 1966. *Woodland Wildflowers of Eastern Saskatchewan.* Regina, Canada, Department of Natural Resources.

——. 1967. *Saskatchewan Wildflowers.* Regina, Canada, Department of Natural Resources.

DENSMORE, F. 1928. *How Indians Use Wild Plants for Food, Medicine and Crafts.* New York, Dover Publications, Inc.

HUTCHENS, A. R. 1971. *Indian Herbology of North America.* Boston/London, Shambhala.

——. 1992. *A Handbook of Native American Herbs.* Boston/London, Shambhala.

JAEGAR, E. 1987. *Indian Reprints: Wildwood Wisdom.* New York, The Macmillan Company.

JASON, D.; JASON, N.; GILBERT, L. 1975. *Some Useful Wild Plants.* Vancouver, Talon Books.

KEANE, K.; HAWARTH, D. 1995*a*. *Herbal Medicine Making.* Saskatchewan, Root Woman & Dave.

——. 1995*b*. *The Native Garden.* Saskatchewan, Root Woman & Dave.

KEANE, K. 1992. *Useful Wild Plants of Saskatchewan. Book 1.* Saskatchewan, Root Woman & Dave.

——. 1993. *More Useful Wild Plants of Saskatchewan. Book 2.* Saskatchewan, Root Woman & Dave.

——. 1994. *Native Medicines.* Saskatchewan, Root Woman & Dave.

LEIGHTON, A. 1986. *A Guide to 20 Plants and Their Uses by the Cree.* La Ronge, Canada, Education Branch of the Lac La Ronge Indian Band.

NEITHAMMER, C. 1974. *American Indian Food and Lore.* New York, Collier Books.

ROWE, J. S.; TEED, L. M. n.d. *Saskatchewan Trees.* Saskatoon, Canada, Tri-Leaf Publications.

Saskatchewan Environmental Resources Management. 1993. *Guide to Forest Understory Vegetation in Saskatchewan.*

SHAY, C. T. 1984. Plants and people: past ethnobotany of the northeastern prairie. *The Prairie: Past, Present, and Future.* Proceedings of the Ninth North American Prairie Conference, July/August. Minnesota.

TREEMENDOUS SASKATCHEWAN FOUNDATION INC. 1992. *TREEmendous Trees and Shrubs.*

VANCE, F. R.; JOWSEY, J. R.; MCLEAN, J. S. 1977. *Wildflowers Across the Prairies.* Saskatoon, Canada, Western Producer Prairie Books.

WEINER, M. 1972. *Earth Medicines – Earth Foods.* New York, Collier-Macmillan Publishers.

Audiotapes

NERBURN, K.; MENGELKOCH, L. (eds.). 1993. *Native American Wisdom.* California, New World Library. (audio tape)

Resource People

Brass, C. Prince Albert, Saskatchewan.

Isbister, Rev. J. Ahtahkakoop First Nations.

Milne, S. Lac La Ronge Indian Band. La Ronge, Saskatchewan.

Wilson, V. Elder, Wonska Cultural School. Prince Albert, Saskatchewan.

Duties to implement reproductive rights: the case of adolescents

Rebecca Cook* and Rika Pretorius**

The significance of applying human rights to advance reproductive health has gained recognition and momentum through recent United Nations conferences, particularly the International Conference on Population and Development (Cairo, 1994) and the Fourth World Conference on Women (Beijing, 1995). The 'Programme of Action' adopted in Cairo made the following observation about reproductive health:

[reproductive health is] a state of complete physical, mental and social well-being and is not merely the absence of disease or infirmity, in all matters relating to the reproductive system and to its functions and processes. Reproductive health therefore implies that people are able to have a satisfying and safe sex life and that they have the capability to

* Director, International Human Rights Law Program, Faculty of Law, University of Toronto, Canada.

** Formerly of the Department of Constitutional and Public International Law, University of South Africa. Since completion of this chapter, Rika Pretorius has died. The chapter is published in her memory to advance the protection and promotion of reproductive health in South Africa, a cause to which she was profoundly committed.

reproduce and the freedom to decide if, when and how often to do so (United Nations, 1994*b*).

The Beijing 'Declaration' and the Beijing 'Platform of Action' explain that 'reproductive rights embrace certain human rights that are already recognized in national laws, international human rights documents and other consensus documents' (United Nations, 1995*b*).

Human rights relevant to reproductive choice have progressively become defined upon the foundation established in 1948 by the Universal Declaration of Human Rights (United Nations, 1948). The Declaration itself was not proposed by the United Nations as a legal instrument, but it has gained legal acceptance and enforceability through a series of international human-rights conventions. The Declaration's two initial legally-binding, implementing covenants are the International Covenant on Civil and Political Rights and the International Covenant on Economic, Social and Cultural Rights.

Human sexual reproduction obviously involves both sexes, but its burden and potential for oppression fall primarily on women. The modern human-rights treaty specifically addressing women's rights is the Convention on the Elimination of All Forms of Discrimination Against Women. The Convention on the Rights of the Child reinforces the Women's Convention by committing states to prohibit discrimination against children, particularly girls. These conventions give more detailed expression to the values implicit in the Universal Declaration of Human Rights.

Similarly derived from this Universal Declaration are regional human-rights conventions holding legal force, including the European Convention for the Protection of Human Rights and Fundamental Freedoms, the American Convention on Human Rights and the African Charter on Human and Peoples' Rights. Of particular significance to the legal prohibition of violence against women, including sexual abuse, is the Inter-American Convention on the Prevention, Punishment, and Eradication of

Violence Against Women (known as the Convention of Bélem do Pará).

The work of non-governmental organizations such as the Commonwealth Medical Association and the International Planned Parenthood Federation, as well as that of broad coalitions of grass-roots women's groups, has significantly improved our understanding of how human rights can be applied to protect and promote reproductive health and self-determination. Still needed to advance reproductive and sexual health is an improved use of national, regional and international means to implement duties so as to respect, protect and fulfil rights. Neither the Cairo nor Beijing documents contain procedures to hold governments legally accountable. However, procedures that can be applied to advance reproductive health and self-determination do exist in national constitutions as well as in international and regional human-rights treaties.

Emerging analyses of state responsibility for violations of human rights assess governmental neglect of preventable causes of reproductive ill health within the larger social framework of systemic unlawful discrimination against women (Cook and Oosterveld, 1995). The content and meaning of national and international human-rights laws have yet to be sufficiently applied to reproductive-health matters. If international human-rights law is to be truly universal, it must be applied with two goals: first, to require states to take preventive and curative measures with regard to reproductive health; and, secondly, to provide individuals with the opportunity to achieve their reproductive security and self-determination (Cook and Fathalla, 1996).

States' duties to implement reproductive rights

A useful framework has been developed under international human-rights conventions, called the Maastricht Guidelines on Violations of Economic, Social and Cultural Rights, to guide and assess the conduct of states that have committed themselves to observing such conventions. They are obligated to undertake three kinds of duties: (1) the duty to respect rights, which prohibits states from interfering with the protection and promotion of reproductive rights; (2) the duty to protect rights, which requires states to prevent conduct by third parties that results in violations of reproductive rights; (3) and the duty to fulfil rights, which requires states to take appropriate legislative, administrative, budgetary, judicial and other measures toward the realization of reproductive rights. These duties require states to take a variety of steps to prevent, remedy and punish violations of reproductive rights.

The duty to respect the rights of adolescents

Respect for the reproductive health rights of adolescents is often denied by state laws and by practices designed to protect adolescents from premature sexual behaviour, the risks of pregnancy and exposure to sexually transmitted diseases. It is acceptable for the state to reinforce parental interests in the protection of vulnerable children. However, when the state aligns itself with parental interests, it may fail to give due respect to adolescents in the struggle between parental protectionism and adolescent autonomy that often marks the developmental stage described as adolescence. Parents are frequently unrealistic in denying their own children's sexuality and capacity for self-determination. Legislators frequently defer to parental demands for reinforcement of their authority and discount the interests of adolescents, who are too young to vote.

National laws designed to protect adolescents – for instance, by denying them access to information and means of contraception, by conditioning their reproductive health care on the consent of parents with whom they may be in conflict over sexual matters or by incriminating their voluntary sexual behaviour – not only fail in their protective purpose, but may violate international human rights (Packer, 1997). The conviction that parents will educate their children at home about sexuality is frequently false, and schools may feel inhibited in conducting sex education programmes for fear of violating parents' interests and laws against children's involvement in sexual activities.

The Cairo and Beijing documents urge governments to address adolescent sexuality through accessible and comprehensible educational programmes in sexual and reproductive health and to provide counselling services on contraception and sexually transmitted diseases. Information in school curricula can be controversial, because sexual biology and behaviour may be explained in ways that parents oppose, at a time they consider premature, or with the effect of causing children to ask questions at home with which parents are uncomfortable. The European Court of Human Rights has respected sensitivity to parents' views, but has also upheld a compulsory sex education course in Danish schools. The court decision states that the 'the curriculum is conveyed in an objective, critical and pluralistic manner [and does not] pursue an aim of indoctrination that might be considered as not respecting parents' religious and philosophical convictions' (*Kjeldsen* v. *Denmark*, 1976).

The Cairo and Beijing documents encourage educational systems that eliminate all barriers impeding the schooling of married and/or pregnant girls and young mothers. Such a barrier was removed in 1995, for instance, when the Botswana Court of Appeal ruled unconstitutional a college regulation that required female students to inform the college director if they were pregnant, rendering them liable to suspension or expulsion (*Molepolole College of Education* v. *Attorney General of Botswana*, 1995).

One of the reasons why adolescents do not use the resources that states are legally obliged to provide to protect their reproductive health is that they do not trust health care professionals to maintain confidentiality. They fear that the disclosures they make relating to their sexual behaviour, disclosures necessary for appropriate health care, will in turn be related to their parents, to the parents of their partners, to schoolteachers or other adults. The revealing of information is feared to be either direct or indirect, as when adolescent health care is covered by schemes of payment involving their parents. Accordingly, the provision of reproductive health care may be useless unless it incorporates credible provisions against a breach of confidentiality.

The legal duty to ensure confidentiality may be founded on, among other things, the provisions in the Convention on the Rights of the Child. Article 16 protects children against interference with privacy, and Article 14(2) requires respect for freedom of thought and conscience, respecting parental rights 'in a manner consistent with the evolving capacities of the child'. This means that health care providers must take into account adolescents' evolving capacities for sexual responsibility.

The duty to protect the rights of adolescents

The duty to protect the rights of adolescents requires that state agencies and their employees try to prevent violations on adolescents' reproductive-health rights by private individuals or organizations not directly bound by international human-rights laws, although perhaps bound by national laws that incorporate international rights. States remain directly bound to give protection against such violations, even when, on grounds of public economy, they transfer what were formerly state functions to private individuals and agencies. Their continuing duty under international human-rights law is 'to organize the governmental apparatus and, in general, all the structures through which public power is exercised so that they are capable of juridically ensuring the free and full enjoyment of human rights' (*Velasquez Rodriguez Case*, 1988).

For example, in a case involving the rape of a mentally impaired girl, the government of the Netherlands denied liability for her consequent physical and mental distress (*X and Y* v. *The Netherlands*, 1985). The European Court of Human Rights maintained, however, that the state had 'a degree of responsibility' for these impairments to her health. The World Health Organization describes health as 'a state of . . . physical, mental and social well-being' (WHO, 1946).[1] The question of liability arose because no means were provided for the assailant to be prosecuted or for the victim to be compensated for injury to her health. The European Court of Human Rights held that the state was required to take positive measures concerning events that occurred between private individuals where human rights had been violated, concluding that the state had failed in its legal duty to protect the rights of adolescent victims by imposing sanctions upon individual violators and by attempting to deter potential violators.

The protection of adolescents requires that they be spoken to and informed about all aspects of sexuality. However, the adults who are influential in their lives, such as parents, teachers, religious leaders and health-care practitioners, may lack the capacity to discuss sexual matters in a language adapted to them or they may simply lack the capacity to do so entirely. Judgmental language prohibiting adolescent curiosity about sex and sexual experimentation is inadequate. For example, during a debate on abortion in South Africa, a feminist activist, Nomboniso Gasa, wrote an open letter to members of Parliament in 1996 arguing for reproductive self-determination. In this letter she relates how she had first heard the Xhosa word meaning 'to abort', *ukuqhomfa*, when she was seven years old:

1. Par. 72 of the *Report of the International Conference on Population and Development* (United Nations, 1994b), quoted at the beginning of this chapter, does not cite this WHO document although it uses the exact wording thereof.

throughout my childhood and as a young woman, we heard of people who had aborted, we saw foetuses and sometimes fully developed babies in the rivers, in the dongas and where women went to collect firewood. In answering my nagging questions, my dad snapped and said, 'Once and for all let me answer you, and after this just shut up! Abortion is real, unpleasant, frightening and it happens. But, I do not want to talk about it, we do not talk about it – people do not' (*Mail and Guardian*, 1996).

Silence about abortion, contraception, sexuality and the ways to prevent sexually transmitted diseases is pervasive not only among parents, but also among teachers of adolescents, government agencies and national and international health protection agencies. Unless they work together to break hypocritical silence regarding adolescent sexuality, the protection of adolescents' rights to reproductive health will be violated.

A common fear is that educating adolescents about sexuality will trigger their curiosity and experimentation. However, sexuality is not a secret to adolescents; it is part of their daily experiences, whether these be in an urban or rural setting, as well as in peer-group discussions. Indeed, such discussions are often adolescents' primary sources of sexual information, and particularly of misinformation, which more responsible education could correct. Governments have a critical duty to ensure adolescents' access to the information necessary for the protection of their reproductive health, as well as a related duty to remove legal, regulatory and social barriers to essential information and care.

The duty to uphold the rights of adolescents

The duty to uphold the rights of adolescents requires that states take appropriate legislative, administrative, judicial, budgetary, economic and other measures to achieve adolescents' full realization of their human rights. Failure on the part of governments to address the magnitude of violations of adolescents' reproductive rights places the state in breach of its duty. For example, while states may have laws that prohibit sexual intercourse with girls below a given 'age of consent', they should not have laws that obstruct girls younger than the age of consent, or younger than any particular age, from exercising a competent capacity to consent to medical care in general, and reproductive health care in particular. Many legal traditions recognize that adolescents below a legal age of consent may be mature minors who are able to make many important decisions in their lives as if they were adults. Laws may proscribe specific limitations on age for such purposes as marriage without parental consent, marriage subject to parental consent, compulsory schooling and eligibility to hold a driver's licence. However, controlling a competent adolescents' access to counselling and health services by subjecting their requests to parental notification and veto can be construed as a violation of their human rights.

States should ensure that health-care providers recognize their duty to assess whether adolescents who request assistance in reproductive health care will be capable of exercising reasonable judgement. Furthermore, they must recognize their legal duty to provide treatment to those who act as adults regarding the issues of the treatment to be provided and confidentiality. One sign of maturity in minors is their understanding of the need to protect their reproductive health, indicated by their requesting contraceptive services when they are, or are about to become, sexually active. A general rule is that adolescents capable of freely choosing to be sexually active without parental control are equally capable of receiving reproductive health counselling and care without parental control.

Enforcing the reproductive rights of adolescents

Means of accountability

The means of accountability that may indicate violations of adolescents' reproductive rights exist at national and international levels. Legal means to make governments accountable are designed to expose governments when they fail to assure their legal duties, to indicate by what standards and processes governments may conform to their legal duties, and to show their compliance with the responsibilities they have assumed. The primary role of national and international human-rights agencies is not punishment, but support of states' intentions of compliance.

Governments may demonstrate to their own satisfaction that they are meeting their duties, but experience shows that there must be international scrutiny of countries' observance of their commitments under international human-rights law. For instance, through human-rights treaties such as the Women's Convention and the Children's Convention, states have committed themselves to report regularly to the respective treaty-monitoring committees on their activities to protect and promote the reproductive rights of adolescents. Monitoring committees are mandated to be vigilant in their scrutiny of states' reports. The Committee on the Elimination of Discrimination Against Women (CEDAW) therefore receives alternative reports or comments on state performance submitted by national and international non-governmental organizations, which may incorporate significant findings of failures to prevent, remedy and punish violations of adolescents' reproductive rights.

To assist countries in their reporting obligations, CEDAW and other such organizations have developed a series of General Recommendations or General Comments, which identify precisely what state reports should address (1996*b*). For instance, in CEDAW General Recommendation Number 14, on the issue of excision, states are urged to 'take appropriate and effective measures with a view to eradicating the practice of female circumcision', including, for instance, the dissemination of information, the provision of educational and training programmes and the support for women's organizations working to eliminate harmful traditional practices.

An increasingly important means for developing state accountability is the publication by treaty-monitoring bodies of Concluding Observations or Concluding Comments on states' reports. For example, in considering the report of the United Kingdom in 1995, the Committee on the Rights of the Child expressed concern that parents could unilaterally 'withdraw their children from parts of the sex education programmes in schools' (1995*d*). Regarding the report of Paraguay in 1997, the Committee noted its concern for 'the absence of large-scale public campaigns for the prevention of unwanted pregnancies, STDs and HIV/AIDS, especially for children and adolescents' and suggested that Paraguay 'promote adolescent health by strengthening reproductive-health and family-planning services to prevent and combat HIV/AIDS, other STDs and teenage pregnancy' (1997*a*). In 1997 CEDAW condemned St. Vincent and the Grenadines for its 'very high rate of pre-teen and teenage pregnancy' and recommended improved reproductive and sexual health services, including sexual education and family planning (1997*b*). Also in 1997 CEDAW expressed grave concern that in Namibia 'pregnant [teenagers] were punished by expulsion from school' (1997*c*).

In Peru, the medical profession is required to facilitate women's access to safe abortions and related health services as the law permits. Moreover, since prevailing laws were shown to result in inhumane treatment of women and undue maternal mortality, Peru is obliged to consider law reform in compliance with human-rights standards for women's health and dignity.

Mechanisms exist under some conventions, such as the European Convention, the American Convention, the Convention of Bélem do Pará and the Political Covenant that enable private individuals from consenting states to bring individual complaints against their states for violations. The move to allow individual petitions is an issue that is currently under consideration for the Women's Convention. A normal condition of international tribunals receiving individuals' petitions, however, is that such individuals exhaust all reasonable possibilities of achieving remedies before national tribunals of the state against which the petition is presented.

The United Nations Special Rapporteur on Violence Against Women receives communications about alleged incidents of gender-specific violence against women that have not been effectively addressed through national legal systems. The Rapporteur enters into dialogue with governments about investigating the merits of alleged violations with a view to reaching resolutions, and reports annually to the United Nations Commission on Human Rights on causes and consequences of violence against women, including violence against girls, implicating their reproductive rights.

Categories of violations

Violations of rights consist of failures to observe legally binding duties. Such violations have been divided into three categories (Chapman, 1996), which may be illustrated with reference to rights relating to adolescent reproductive health:

- category 1 violations result from direct actions on the part of states, such as state interference with adolescents' access to information;
- category 2 violations relate to the failure of states to meet the minimum core obligations of human rights, such as neglecting to have and/or enforce laws prohibiting excision;

- category 3 violations consist of patterns of discrimination, such as persistent and gross discrepancies in access to health services, which cumulatively disadvantage the reproductive health of groups, such as adolescents.

Discrimination on grounds of age is comprehensively addressed through the children's Convention. States agree 'to ensure that no child is deprived of his or her right of access to . . . health care services' (Convention on the Rights of the Child, 1989). However, the Cairo Programme recognizes that the 'reproductive health needs of adolescents as a group have been largely ignored to date by existing reproductive health services' (1994*b*). As a result, unmarried pregnancy in adolescence has now reached epidemic proportions in many countries, while in other countries, high levels appear endemic (Senanayake and Ladjali, 1994).

The Cairo and Beijing documents call for the removal of regulatory and social barriers to reproductive health information and care for adolescents. They recommend that countries ensure that the programmes and attitudes of health-care providers not restrict adolescents' access to appropriate services, but rather protect and promote their rights to reproductive-health education, information and care in order to reduce the number of adolescent pregnancies.

Effectiveness of remedies

It has been universally recognized that if discussion about rights is not accompanied by discussion about solutions to the injustices resulting from 'rights' violations, then the discussions may remain severely limited. It is generally recognized that legal systems have a duty to ensure that solutions or remedies to existing injustices be accessible, justly administered and effective. Addressing cases of breaches of women's human rights, particularly at the international level, does not primarily involve imposing sanctions, but rather ensuring that violations of rights will be prevented or deterred in the future. As has already been stated, human-rights agencies play a supportive, not a punitive, role.

With a greater understanding of why violations of adolescents' reproductive rights occur, legal systems will be better equipped to ensure the effectiveness of remedies. For example, research into high rates of adolescent pregnancy can show how effective remedies may be formulated. A study conducted in the state of Washington (United States) showed that two-thirds of a sample of 535 pregnant young women became pregnant as a result of sexual abuse by older men (Boyer and Fine, 1992). These findings could be used to encourage governments to address the problem of sexual abuse of adolescent girls, in addition to their providing contraceptive services. In other words, the research demonstrates that many pregnancies were due not to any failure on the part of the girls, but to failures to protect them against abuse. Enforcement of domestic statutory rape laws may be a means to punish those men who prey on young girls, although there is some doubt that such laws will have an effect on adolescent sexual activity and pregnancy rates (Donovan, 1997). Most certainly, the enforcement of statutory rape laws should not replace the provision of reproductive health information and services to adolescents.

The way forward

Until recently, reproductive-health concerns, including those of adolescents, were neglected by international human-rights law. However, the United Nations conferences in Cairo and Beijing, along with enlightened comments and observations by international treaty bodies, such as the Children's Rights Committee and CEDAW, have begun to direct international attention to the need to implement the reproductive rights of adults and adolescents alike. The wider health and welfare of adolescents and the principles of social justice demand that governments bring their laws, practices and policies into line with the standards outlined in the Cairo and Beijing documents. Under regional and international human-rights treaties, such as the children's convention and the women's convention, states must not only not interfere with the exercise of such rights, but must also encourage third parties not to interfere with them, and even take positive steps to ensure that adolescents have the information and access to the services they require.

It remains the concern of national governments, concerned NGOs and medical groups to bring laws, policies and practices into compliance with human-rights obligations. Experience has shown that failure to address the reproductive-health concerns of adolescents will not delay sexual activity or prevent teenage pregnancy. Rather, failure to recognize the basic human rights of adolescents puts the health, and even the lives, of young people at an unacceptable risk.

Summary

Lorraine Dennerstein*

All life, especially human life, involves ethical questions. Concern about health ethics developed as part of wider international concern for human rights following the Second World War. Because bioethics deals with the beliefs, values and norms basic to a society, its cultural tradition and its collective conscience (Fox,1992), bioethical views must be understood in the context of the culture in which they arise; moreover, they are widely influenced by societal change and political and religious movements. Observable behaviour is complex and driven by many factors and cannot be understood without consideration of the social context and political frameworks. For example, governments do not use principles of distributive justice in allocating resources to health care because government policy is located within politics rather than bioethics (Lewins, 1996). An important issue in feminist approaches to bioethics is the influence of gender or the social construct of what it means to be male or female. This approach questions the dominance of male western

* Director, Office for Gender and Health, Department of Psychiatry, University of Melbourne (Australia).

approaches to bioethics which presume that the moral subject is an autonomous, detached, rational one. Feminist writers such as Gilligan (1982) question whether judgements can be made independently of the human emotions involved, of the situation or context and of the dynamics of relationships. Other feminist scholars such as Wolf (1996) have given possible reasons for why bioethics has remained seemingly insensitive to the effects of gender for so long. Wolf points to the focus of bioethics on individualism, and inattention to the moral significance of groups and the social context; emphasis on deduction from ethical principles, rather than examination of concrete cases; and the failure of bioethics to be self-critical about who the field serves and how.

Some bioethical principles, like autonomy, which implies that an individual is able to fix a course of action voluntarily, may not be invokable in societies in which women have no property rights or cannot make independent decisions about their own health. Even in some developed countries in which legislation has ensured the right to autonomy, women are unable to exercise that right in decisions about health care if they are not provided with adequate and understandable information which enables such decision making. Autonomy is linked to justice, since both are concerned with treating people with equality regardless of social attributes such as gender, age or religion. Clearly, issues of justice cannot be considered in isolation from the social context which may include a gender bias. Even societies whose legislation states that there should be equal opportunity for both sexes may have customary practices and other laws that do not reflect gender equality.

The past two decades have seen growing global recognition that there are pronounced gender differences in health and that women's health disadvantage is clearly linked to the continuing socio-economic and political disadvantages of women in most societies. International concerns about women's health and rights have been widely discussed this decade and received considerable impetus from four United Nations conferences (Human Rights, Vienna 1993; International Conference on Population and Development, Cairo 1994; World Summit for Social Development, Copenhagen 1995; and the Fourth World Conference on Women, Beijing 1995). The universality of women's human rights was affirmed. We have highlighted in this book some areas of substantial disadvantage in women's health outcomes involving women's position in society where bioethical and human rights issues must be addressed. Discrimination against women occurs despite the Universal Declaration of Human Rights (1948) and treaties specifically addressing women's rights: the Convention on the Elimination of all Forms of Discrimination Against Women and the Convention on the Rights of the Child.

Bioethical debate, or debate about the ethics of health care, has often been driven by the rapid advancement of technology. Developing an internationally agreed bioethical statement for the human genome was a priority for the International Bioethics Committee of UNESCO (UNESCO, 1997). Advanced reproductive technologies such as genetic screening, surrogacy, frozen embryos and fetal tissue remain at the centre of public interest, debate and legislation. It could be argued that the amount of academic, media and public debate on such issues has been out of proportion when the percentage of those affected is considered. These areas of reproductive technology necessarily involve and impact on women and men. Yet the proponents of bioethical debate and those involved in utilizing these technologies have often not considered how gender effects technology development and implementation, or even the reality that the person whose body is at stake is a woman (Wolf, 1996). While certain reproductive technologies (ultrasound and genotyping) were developed to detect congenital abnormalities during pregnancy – thus to help women, and men – their use may cause ethical dilemmas, such as whether to abort a malformed or genetically 'abnormal' fetus. When the outcome is sex-selective abortion, based solely on the sex (usually female) of the fetus, bioethicists must be concerned about the use of modern technology against the female foetus.

Other gender related areas which continue to cause immense and needless morbidity and mortality are the health issues embedded in social values such as those revolving around access to the health care resources needed for reproductive health and safe motherhood. Pregnancy-related factors cause the deaths of nearly half a million women each year and vary with the development status of many countries. Most of these deaths could easily be prevented if women were given access to adequate health care. Unsafe abortions threaten the lives of a large number of women. A further 30 million women among those who become pregnant each year will suffer severe and long-term disabilities. STDs disproportionately affect women. Social vulnerability and unequal power relationships between women and men are major obstacles to 'safe' sex. Yet the role of the male in these and other issues impacting on the health of women and children has often been ignored by medical research and practice.

Some traditional practices, such as excision, are variously estimated to affect as many as 130 million women worldwide. Such practices, based on the desire to subjugate female sexuality, involve the removal of healthy, functional body tissue and have serious reproductive health consequences. Thus the ethical duty to non-maleficence is invoked, but the practice can be understood and solutions found only by understanding the role of gender in social and sexual relationships in those societies which carry out such practices.

Differential treatment of the female infant occurs in many countries, with sex selective abortion, infanticide, shorter breast feeding, less secondary feeding and less preventive health care resulting in higher mortality, inadequate nutrition and poor health. UNDP estimates at one million the number of excess female infant deaths annually (UNDP, 1995). This occurs despite a sounder biological constitution of the female infant and the Convention on the Rights of the Child. Even when education is guaranteed by the constitution of a country, many girls, particularly in rural areas, are denied education or given fewer years of education than boys. Differential access to education has lifetime implications on family size and on women's ability to provide health care for themselves and their families, and on their economic status and autonomy. It is apparent that the prevalence of these problems varies across and within different nations and is linked to socio-economic development, educational level and the value ascribed to being female in that society.

Violence against women, which often takes place in the home, has only recently been recognized as a major cause of physical and mental ill health and is probably underestimated. Gender-based violence includes ill treatment of girls, incest, rape, battery, homicide, psychological abuse, forced prostitution, trafficking of women, sexual harassment, genital mutilation, and dowry-related murder. Rape during war has only recently been publicly acknowledged as a war crime. Gender-based violence continues to affect women in nearly all societies despite the Declaration on the Elimination of Violence Against Women.

Depression is a major contributor to health impairment worldwide, with higher impact and prevalence in women. Mental disorders among women are often traced to marginalization, powerlessness, poverty, overwork and stress, domestic violence and substance abuse which often reflect gender inequity.

Environmental risks in the home and workplace may have a disproportionate impact on women's health because of their different susceptibilities to toxic effects of various chemicals. Occupational health issues are of increasing importance with many women in the informal sector where there is no regulation of conditions (Dennerstein and Kane, 1999).

Women form the majority of the elderly in most countries. Older women need particular protection of their human rights if they are to maintain a decent level of health. This particularly applies to those societies where widows are denied property rights, or where gender has played a role in unequal access to economic resources which may be needed to maintain nutrition, safe housing and health care later in life. New burdens of caring for grandchildren have been placed on older women with internal migration of children and the effects of AIDS-related deaths of the parents. Changing family structures and social roles have also left the elderly exposed to a lack of family care and protection. Abuse of the elderly has been reported in many different countries of varying economic status.

Gender is not the only adverse factor affecting health. Certain social factors such as race, class and poverty have large effects and there are complex relationships between these factors and gender. Thus there is a need for an interdisciplinary and inter-sectoral approach to bring together knowledge and combine insights and strategies from a range of disciplines, such as health sciences, social sciences, education, engineering, law, and economics. It must be informed by community involvement. Bioethics could play a major role in developing such an approach.

Some reflections in conclusion

Michèle Jean*

Increasing attention is being paid to the relationship between health, human rights and bioethics. Several authors even speak of a new generation of human rights and are putting forward the idea of a new convention or a new charter of basic rights. Current scientific research, especially in genetics, is helping to fuel these debates. A large proportion of these discussions relates more specifically to women. It is a healthy sign that, everywhere in the world, women are concerned with these discussions, asking questions and seeking to participate in a sustained manner in the redefinition and definition of the paradigms that demarcate or are going to demarcate the contours of these questions and their answers.

When discussing, *inter alia*, patients' rights, the right of access to health care, informed consent, the confidentiality of data and the status of the embryo, it is important that the aspects relating more specifically to women be taken into account in every sector and in every dimension being studied. It is essential, and

* Historian, Under-Minister of Health (Canada) from June 1993 to July 1998.

the different chapters of this book demonstrate this point, that women in the future should be closely associated with the preparation of the policies and documents that are bound to be formulated in years to come. History has shown that a lack of participation by certain social groups in the definition of cultural, social and economic policies leads to decisions and forms of behaviour that are unjust and discriminatory towards these groups. Today, one would speak of unethical behaviour and decisions.

The various chapters of this book identify certain major trends in socio-economic and cultural sectors that affect certain modes of action and thought over time, and that have been detrimental to the equal participation of women in society, especially as regards their health.[1] At the ethical level this entails a number of ethical principles to be respected at every stage in the life of women: before birth and during childhood, puberty and adolescence, adult life and old age. The following list gives the principles which should form the basis of any discussion on women's rights and bioethics.

- Access to: general health care; reliable services for the termination of pregnancy; clinical trials and research protocols; knowledge through formal and informal education, higher education and research groups; responsibility through community, local, regional and provincial decision-making structures; and appropriate information conveyed by traditional means of communication as well as by means of new technologies.
- In respect of human dignity and justice, legislation must recognize the need to respect women's dignity and freedom of choice; the exactions perpetrated in times of war must be regarded as war crimes and must be punished; young girls and teenagers must be protected from sex tourism and

1. For an interesting feature article on the question of women's health, see 'The truth about women's bodies', *Time*, 15 March 1999, pp. 40–51.

prostitution; and marital violence must be deterred and the necessary means set up to help women and men in distress.

- Considering the right to life, all policies and intervention on genetic material likely to lead to gender selection must be examined with the utmost caution in order to detect any ill effects and prohibit them if necessary; and measures must be implemented to ensure the right to proper nourishment.
- As regards non-discrimination in genetic testing, specifically female genetic characteristics should not be used to discriminate against women. In national statistics, data (such as epidemiological data, etc.) should be collected everywhere according to sex so as to enable the formulation and adjustment of policies concerning women.

These are only some of the dimensions that need attention when examining the problem of women's health in the context of bioethics and human rights. Far too often, women have had to react after the event to the issues that concern them. Today we have knowledge, especially in fields involving women's sexual functions. We should profit from these discussions by not hesitating to challenged received ideas. As Jean Lacroix said, 'It is necessary to be contemporary in one's own thinking, that is to say, to constantly rebuild all one's knowledge.'[2]

From a historical perspective, it can be asserted that the study of women's condition is recent. We must continue, in spite of difficulties and obstacles, to develop this field of study because, as has already been said, the quality of a society can be measured by the way in which it treats its women.

2. 'Une pédagogie de la raison', *Le Monde*, 27 March 1999, p. 16.

Bibliography

ACHESON, D. 1993. Health Humanitarian Relief and Survival in Former Yugoslavia. *Br. Med. J.*, Vol. 307, pp. 44–8.

ADNAN, S. 1993. Birds in a Cage: Institutional Change and Women's Position in Bangladesh. In: N. Federici, K. Oppenheim Mason, and S. Sogner (eds.), *Women's Position and Demographic Change*. Oxford, Clarendon Press, pp. 285–318.

ANJALI. 1987. Prejudice Against Girls: Abuse of Scientific Discovery. *Social Welfare*, Vol. 34, pp. 10–11.

ARCEL, L. T.; FOLNEGOVIC-SMALC, V.; KOZARIC-KOVACIC, D.; MARUSIC, A. 1995. *Psychosocial Help to War Victims: Women Refugees and their Families*. Copenhagen/Zagreb, International Rehabilitation Council for Torture Victims.

ARIAS LONDOÑO, M. 1991. *Cinco Formas de Violencia contra la Mujer*. Bogotá, Ecoe Ediciones.

ARMSTRONG, B. 1986. Involving Young Men in Family Planning Services. *Planned Parenthood Review*, Autumn, pp. 4–6.

AUSTRALIAN INSTITUTE OF HEALTH. 1990. *Australia's Health 1990*. 2nd Biennial Report of the Australian Institute of Health. Canberra, Australian Government Publishing Service.

AVIS, N. E.; MCKINLAY, S. M. 1991. A Longitudinal Analysis of Women's Attitudes towards Menopause: Results from the Massachusetts Women's Health Study. *Maturitas,* Vol. 13, pp. 65–79.

BALASUBRAMANYAN, V. 1986. No Girls Please, We're Indian. *Mainstream,* March 1, p. 32.

BALTES, M. M.; HORGAS, A. 1998. Ageing and Mental Health. In: H. Friedman (ed.), *Encyclopedia of Mental Health.* San Diego, Academic Press.

BALTES, P. B.; MAYER, K. U. (eds.). 1999. *The Berlin Aging Study.* New York, Cambridge University Press.

BELLANTONI, M. F.; BLACKMAN, M. R. 1995. Menopause and its Consequences. In: E. L. Schneider and J. W. Rowe (eds.), *Handbook of the Biology of Aging.* New York, Academic Press, pp. 415–30.

BLOOR, M. 1995. *The Sociology of HIV Transmission.* Thousand Oaks, California, Sage Publications.

BOLVARY, K.; VACZIZ, M. 1996. Four Year's Experiences of the AIDS Outreach Program for Sex Workers in Budapest. Paper to the 11th International Conference on AIDS. Vancouver, Canada.

BONITA, R. 1996. *Women, Ageing and Health: Achieving Health Across the Life Span.* World Health Organization, Geneva.

BORCHERS, A. 1986. Born to Die. *India Today,* June.

——. 1997. *Die Sandwich Generation.* Frankfurt, Campus.

BOSE, A. 1985. Boy or Girl: Choose Your Child Through Diet. *Stardust* (Bombay), July.

——. 1991. *Population of India: 1991 Census Results and Methodology.* New Delhi, BR Publishing.

BOYER, D.; FINE, D. 1992. Sexual Abuse as a Factor in Adolescent Pregnancy and Child Maltreatment. *Family Planning Perspectives,* Vol. 24, pp. 4–11.

BRACKEN, P. J.; GILLER, J. E.; KABAGANDA, S. 1992. Helping Victims of Violence in Uganda. *Medicine and War,* Vol. 8, pp. 155–63.

BROWNMILLER, S. 1975. *Against Our Will: Men, Women and Rape.* New York, Simon & Schuster.

BUMILLER, E. 1991. *May You Be the Mother of a Hundred Sons.* Delhi, Penguin Books.

BYLES, J. E.; FELDMAN, S.; MISHRA, G. 1999. For Richer, for Poorer, in Sickness and in Health: Older Widowed Women's Health, Relationships and Financial Security. *Women and Health*, Vol. 29, No. 1, pp. 15–30.

CAMARA DE COMERCIO DE BOGOTA. 1994. *La Prostitución Infantil en el Centro de Bogotá. Un Ensayo de Investigación Social Urbana.* Taller de publicaciones de la Cámara de Comercio de Bogotá.

CHAPMAN, A. 1996. A 'Violations Approach' for Monitoring the International Covenant on Economic, Social and Cultural Rights. *Human Rights Quarterly*, Vol. 18, pp. 23–66.

COMMONWEALTH DEPARTMENT OF COMMUNITY SERVICES AND HEALTH (Australia). 1989. *National Women's Health Policy.* Canberra, Australian Government Publishing Service.

COOK, R. J.; FATHALLA, M. F. 1996. Advancing Reproductive Rights beyond Cairo and Beijing. *International Family Planning Perspectives*, Vol. 22, pp. 115–21.

COOK, R. J.; OOSTERVELD, V. L. 1995. A Select Bibliography of Women's Human Rights. *American University Law Review*, Vol. 44, pp. 1429–71. (Updated periodically on the internet at: http://www.law.utoronto.ca/pubs/h-rghts.htm).

COPELON, R. 1996. The Potential and Challenge of a Human Rights Perspective. *The Right to Live Without Violence.* Women's Health Collection, No. 1. Santiago de Chile, Latin American and Caribbean Women's Health Network.

CULBERTSON, F. M. 1997. Depression and Gender. *Journal of the American Psychological Association*, Vol. 52, pp. 25–31.

DAHL, S.; MUTAPCIC, A.; SCHEI B. 1998. Traumatic Events and Predictive Factors for Post-Traumatic Symptoms in Displaced Bosnian Women in a War Zone. *Journal of Traumatic Stress*, Vol. 11, No. 1, pp. 137–45.

DAHL, S. 1993. *Rape: A Hazard to Health.* Oslo, Scandinavian University Press.

DEFENSORIA DEL PUEBLO. 1995*a*. *Situación de la Niñez Explotada Sexualmente en Colombia*. Serie Fémina, No. 6. Bogotá, Colombia, Defensoria del Pueblo.

——. 1995*b*. *Avances en la Construccion Juridica de la Iqualdad para las Mujeres Colombianas*. Serie Fémina, No. 4. Bogotá, Colombia, Defensoria del Pueblo.

DENNERSTEIN, L.; ASTBURY, J.; MORSE, C. 1993. *Psychosocial and Mental Aspects of Women's Health*. Geneva, WHO.

DENNERSTEIN, L. et al. 1989. Practice Patterns and Family Life: A Survey of Melbourne Medical Graduates. *Medical Journal of Australia*, Vol. 151, October 2, pp. 386–90.

DENNERSTEIN L.; KANE, P. 1999. *Women and Occupational Health: Issues and Policy Paper for the Global Commission on Women's Health*. Geneva, World Health Organization. (Doc. WHO/CHS/GCWH/99.1.)

DEPARTAMENTO NACIONAL DE PLANEACION. 1995. *Programa Nacional de Atención Integral a la Población Desplazada por la Violencia*. Consejería presidencial para los derechos humanos, CONPES-Ministerio del Interior, Bogotá.

DESJARLAIS, R. et al. 1995. *World Mental Health: Problems and Priorities in Low-Income Countries*. Oxford, Oxford University Press.

DHARMALINGAM, A. 1994. Old Age Support: Expectations and Experiences in South Indian Villages. *Population Studies*, Vol. 48, no. 1, pp. 5–20.

——.1995. Disappearing Girls. *Asiaweek*, January 3.

DONOVAN, P. 1997. Can Statutory Rape Laws Be Effective in Preventing Adolescent Pregnancy? *Family Planning Perspectives*, Vol. 24, pp. 4–11.

DOWELL, S. 1995. Health and Nutrition Centers for Unaccompanied Refugee Children. *Journal of the American Medical Association*, Vol. 273, pp. 1802–6.

EHRHARDT, A. A. 1996. Editorial: Our View of Adolescent Sexuality – A Focus on Risk Behavior without the Developmental Context. *American Journal of Public Health*, Vol. 86, no. 11, pp. 1523–5.

EL-HAMAMSY, L. S. 1995. *Early Marriage in Two Egyptian Villages*. Cairo, Population Council.

ENLOE, C. 1995. The Globetrotting Sneaker. *Ms Magazine*, March/April, pp. 10–5.

Facets, 1982, Vol. 1, No. 3, August.

FARID, S. 1996. Transitions in Demographic and Health Patterns in the Arab Region. Paper presented at the Arab Regional Population Conference, Cairo 8–12 Dec. 1995. Vol. 1, pp. 435–68. Cairo, CDC Press.

FATHALLA, M. F. 1994. Women's Health: An Overview. *International Journal of Gynecology and Obstetrics*, Vol. 46, pp. 105–18.

FELD, S.; SCHOORLE, J.; TRIBALAT, M. 1996. Aspects Démographiques de la Population. Paper presented at the Arab Regional Population Conference, Cairo, 8–12 Dec. 1995. Vol. 1, pp. 3–8. Cairo, CDC Press.

FINGER, W. 1992. Men and Family Planning. *Network, Family Health International*, Vol. 13, No. 1.

FLACSO (Facultad Latinoamericana de Ciencias Sociales). 1995. *Mujeres Latinoamericanas en Cifras*. Santiago de Chile, Tomo Comparativo.

FOX, R. 1992. *The Sociology of Medicine: A Participant Observer View*. Englewood Cliffs, New Jersey, Prentice Hall.

FREEDMAN, L.; MAINE, D. 1993. Women's Mortality: A Legacy of Neglect. In: J. Gabe and P. Williams (eds.), *Tranquillisers: Social and Psychological and Clinical Perspectives*. London, Tavistock.

FRLJAK, A.; CENGIC, S.; HAUSER, M.; SCHEI, B. 1997. Gynecological Complaints and War Traumas. A Study from Zenica, Bosnia-Herzegovia during the War. *Acta Obstet. Gynecol. Scand.*, Vol. 76, pp. 350–4.

FRUZZETTI, L. M. 1997. *Orphans and Abandonment: Gender Liminality and the Women's Movement in India*. Quebec, World Heritage.

GATZ, M.; HARRIS, J. R.; TURK-CHARLES, S. 1995. The Meaning of Health for Older Women. In: A. L. Stanton and S. J. Gallant (eds.), *The Psychology of Women's Health*. Washington, D.C., American Psychological Association.

GEORGE, L. 1980. *Role Transition in Later Life*. Monterey, California, Brooks Cole.

GILLIGAN, C. 1982. *In a Different Voice: Psychological Theory and Women's Development*. Cambridge, Massachusetts, Harvard University Press.

GLOBAL LINK FOR MIDLIFE AND OLDER WOMEN. 1995. *Look at the World through Older Women's Eyes*. Beijing Special Edition, Washington D.C., American Association of Retired Persons.

GRANT, J. 1995. Women and Health: Overview of Three International Conferences. *Proceedings of the Eleventh Commonwealth Health Ministers Meeting, Capetown, 1995*. Commonwealth Secretariat: London, HMM, Vol. 95, No. 17, pp. 216–32.

GROBBELAAR, J. P.; BATEMAN, E. D. 1991. Hut Lung: A Domestically Acquired Pneumoconiosis of Mixed Aetiology in Rural Women. *Thorax*, Vol. 46, pp. 334–40.

HAVIGHURST, R. J. 1948. *Developmental Tasks and Education*. New York, McKay.

HEISE, L. L. 1994*a*. Gender-Based Violence and Women's Reproductive Health. *International Journal of Gynaecology and Obstetrics*, Vol. 46, No. 2, pp. 221–9.

——. 1994*b*. Violence Against Women: The Missing Agenda. In: M. Koblinski, J. Timyan and J. Gay (eds.), *The Health of Women: A Global Perspective*. Boulder, USA, Westview Press.

HEISE, L. L.; Pitanguy, J.; Germain, A. 1994. *Violence against Women: The Hidden Health Burden*. Washington D.C., World Bank.

HELMCHEN, H., et al. 1998. Mental Diseases in Old Age. In: P. B. Baltes and K. U. Mayer (eds.), *The Berlin Aging Study*. New York, Cambridge University Press.

HENSHAW, S. K. 1990. Induced Abortion: A World Review. *Family Planning Perspectives*, Vol. 22, pp. 76–89.

HOFFMAN, B.; MEYER, I. 1992. Neue Perspektiven fur Frauen auf dem Bau. *Soziale Sicherheit*, Vol. 41, No. 5, pp. 149–52.

HOSKINS, I. 1986. India: Make Sure of Baby Boys. *New Scientist* December 25.

——. 1992. *Gender, Aging and Development: Emerging Issues and Policy Recommendations*. Geneva, Ageing International, United Nations.

Indian Express. 1988. New Delhi, December.

INTERNATIONAL LABOUR ORGANISATION. 1995. Gender Issues in Occupational Safety and Health. *Gender Issues in the World of Work*. Geneva, Office of Special Adviser on Women Workers' Questions, International Labour Organisation.

IYENGER, P. 1992. Girls in Salem are Born to Die. *Times of India*, August 30.

KADDAFI, M. 1990. *The Green Book*. Tripoli, Libyan Arab Jamahiriya, International Centre for Study and Research on the Green Book. (In Arabic).

KANE, P. 1991. *Women's Health: From Womb to Tomb*. London, Macmillan.

KEITH, P. M. 1989. *The Unmarried in Later Life*. New York, Praeger.

KESSLER, R. C.; MCGONAGLE, K. A.; ZHAO, S. 1994. Life-Time and 12-Month Prevalence of DSM-III-R Psychiatric Disorders in the United States: Results from the National Comorbidity Survey. *Archives of General Psychiatry*, Vol. 51, pp. 8–19.

KITTRIE, N. K. 1971. *The Right to be Different: Deviance and Enforced Therapy*. Baltimore, Hopkins Press.

Kjeldsen v. *Denmark*, European Court of Human Rights, Rep. 711 (1976), para. 53.

KLERMAN, G. K.; WEISSMAN, M. M. 1989. Increasing Rates of Depression. *Journal of the American Medical Association*, Vol. 261, pp. 2229–35.

KULKARNI, S. 1986. *Sex Determination Tests and Female Foeticide in the City of Bombay*. Report commissioned by the Secretary to the Government Department of Health and Family Welfare, Government of Maharashtra.

KURZ, K. M.; PRATHER, C. J. 1995. *Improving the Quality of Life of Girls*. New York, AWID, UNICEF.

LARRAIN, S. 1993. *Estudio de Frecuencia de la Violencia Intrafamiliar y la Condición de la Mujer en Chile*. Santiago de Chile, OPS.

LARSEN, E. 1996. The Effect of Different Police Enforcement Policies on the Control of Prostitution. *Canadian Public Policy*, Vol. 21, No. 1, pp. 40–55.

LEWINS, F. W. 1996. *Bioethics for Health Professionals: An Introduction and Critical Approach*. Melbourne, Australia, MacMillan.

Libro Blanco de la Mujer. 1994. Propuestas básicas. Documento COMPES del Departamento Administrativo de Planeacion Nacional. Bogotá, Colombia, Presidencia de la Republica.

LIN, LEAN LIM 1993. The Feminisation of Labour in the Asian-Pacific Rim Countries: from Contributing to Economic Dynamism to Bearing the Brunt of Structural Adjustments. In: N. Ogawa, G. Jones and J. G. Williamson (eds.), *Human Resources in Development Along the Asian-Pacific Rim*. Singapore, Oxford University Press, pp. 176–209.

LINGAM, L. 1991. Sex Detection Tests and Female Foeticide: Discrimination Before Birth. *Indian Journal of Social Work*. Vol. L11, No. 1, pp. 13–19.

LOPATA, H. 1987*a*. *Widows, Volume II. North America*. Durham, NC, Duke University Press.

——. 1987*b*. *Widows, Volume I: The Middle East, Asia and the Pacific*. Durham, NC, Duke University Press.

——. 1996. *Current Widowhood: Myths & Realities*. New York, Sage Publications.

LOPATA, H.; HEINEMANN G. D.; BAUM, J. 1982. Loneliness: Antecedents and Coping Strategies in the Lives of Widows. In: L. A. Peplau, D. Perlman (eds.), *Loneliness: A Sourcebook of Current Theory, Research and Therapy*. Brisbane, Australia, Wiley and Sons Interscience Publication.

LOWMAN, J.; FRASER, L. 1995. *Violence Against Persons Who Prostitute: The Experience in British Columbia*. Canada, Dept of Justice and Solicitor General.

LYNCH, M. E. 1991. Gender Intensification. *Encyclopedia of Adolescence*, Vol. 1. New York, Garland Publishing Co.

——. 1998. Maastricht Guidelines on Violations of Economic, Social, and Cultural Rights. *Human Rights Quarterly*, Vol. 20, No. 2, pp. 691–701.

McGrath, E. et al. 1990. *Women and Depression: Risk Factors and Treatment Issues.* Washington, D.C., American Psychological Association.

——. 1996. *Mail and Guardian*. Oct. 18, 1996. (Internet version: http://wn.apc.org/wmail/issues/961018/NEWS67.html).

Makinwa-Adebusoye, P. 1993. Migration and Female-Headed Households. In: N. Federici, K. Oppenhei-Mason, S. Sogner (eds.), *Women's Position and Demographic Change*. Oxford, Clarendon Press, pp. 319–38.

Malawi National Statistical Office. 1987. *Survey of Handicapped Persons, Malawi 1983*. Zomba, Government Printer.

Markson, E. W. 1994. *Issues Affecting Older Women: Promoting Successful Aging*. Belmont Hills, California, Sage Publications.

Martin-Matthews, A. 1991. *Widowhood in Later Life*. Toronto, Butterworths.

Mayer, K. U.; Baltes, P. B. (eds). 1996. *Die Berliner Altersstudie*. Berlin, Akademie Verlag.

Mesle, F.; Shkolnikov, V. M.; Hertrich, V.; Vallin, J. 1996. *Tendances récentes de la mortalité par cause en Russie, 1965–1994*. Paris, Institut National d'Études Démographiques.

Messing, K.; Chatigny, C.; Courville, J. 1996. L'invisibilité du travail et la division léger/lourd: impact sur la santé et la sécurité du travail. *Objectif Prévention*, Vol. 19, No. 2, pp. 13–16.

Miller, K. 1996. The Effects of State Terrorism and Exile on Indigenous Guatemalan Refugee Children: A Mental Health Assessment and an Analysis of Children's Narratives. *Child Development*, Vol. 67, pp. 89–106.

Molepolole College of Eduction v Attorney General of Botswana (for and on behalf of the Principal of Molepolole College of Education and Permanent Secretary of Ministry of Education). Unreported. Civil Appeal no. 13 of 1994, Misca no. 396 of

1993. Judgement delivered on 31 January 1995. Reported in: E. K. Quansah. Is the Right to Get Pregnant a Fundamental Human Right in Botswana? *Journal of African Law*, Vol. 39, pp. 97–102.

MOORE, S. (ed.). 1997. *Female Genital Cutting: Findings from the Demographic and Health Surveys Programs*. Clayverton, Maryland, Macro International.

MORROW, M. 1995. *Women's Health Profile: Vietnam*. Manila, Regional Office for the Western Pacific, WHO.

MOSS, N. 1992. Child Health Outcomes among Central American Refugees and Immigrants in Belize. *Social Science and Medicine*, Vol. 34, pp. 161–7.

MUHONDWA, E. P. Y. 1996. *The Study of the Effects of Vasectomy Promotion Project on Knowledge, Attitudes and Behaviour among Men in Dar Es Salaam*. New York, Population Council.

MURRAY, C. J. L.; LOPEZ, A. D. 1996. *The Global Burden of Disease and Injury*. Harvard, Harvard University Press.

NALUGODA, F. et al. 1997. HIV Infection in Rural Households, Rakai District, Uganda. *Health Transition Review*, Vol. 7 (suppl. 2), pp. 127–40.

NOLEN-HOKSEMA, S. 1990. *Sex Differences in Depression*. Stanford, California, Stanford University Press.

NUPE (National Union of Public Employees [UK]). 1993. *Bringing It All Home: the NUPE Home Care Survey*. London, NUPE.

O'BRYANT, S. L.; MORGAN, L. A. 1989. Financial Experience and Well-Being among Mature Widowed Women. *The Gerontologist*, Vol. 29, No. 2, pp. 245–51.

OFFICE OF AGEING. Department of Family Services and Aboriginal and Islander Affairs. 1994. *Abuse of Older People in Queensland: A Report of the Project on Abuse of Older People*. Queensland, Australia, Community Services Development.

OFFICE OF THE STATUS OF WOMEN, Australian Bureau of Statistics. 1994. *Australian Women's Yearbook 1994*, pp. 44–5. Canberra. Australian Government Publishing Service.

OMRAN, A. R.; STANDLEY, C. C. 1976. *Family Formation Patterns and Health.* Geneva, World Health Organization.

——. 1981. *Further Studies on Family Formation Patterns and Health.* Geneva, World Health Organization.

ORUBULOYE, I. O.; CALDWELL, P.; CALDWELL, J. C. 1994. The Role of High Risk Occupations in the Spread of AIDS: Truck Drivers and Itinerant Market Women in Nigeria. In: I. O. Orubuloye et al. (eds.), *Sexual Networking and AIDS in Sub-Saharan Africa.* Health Transition Series No. 4, pp. 89–100. Canberra, Australian National University.

ORY, M. G.; WARNER, H. R. (eds.). 1990. *Gender, Health, and Longevity: Multidisciplinary Perspectives.* New York, Springer.

OWEN, M. 1996. *A World of Widows.* London, Zed Books.

PACKER, C. A. A. 1997. Preventing Adolescent Pregnancy: The Protection Offered by International Human Rights Law. *International Journal of Children's Rights*, Vol. 5, pp. 47–76.

PAN, S. 1996. Qualitative Study on the Status Quo of Sex Workers and their AIDS Related Sexual Behaviour in Two Chinese Cities. Paper presented at the 11th International Conference on AIDS, Vancouver.

PARK, CHAI BIN; CHO, NAM-HOON. 1995. Consequences of Son-Preference in a Low-Fertility Society: Imbalance of the Sex Ratio in Korea. *Population and Development Review*, Vol. 21, No. 1, pp. 59–84.

POPULATION REFERENCE BUREAU. 1996*a*. *World Population Data Sheet.* Washington, D.C.

——. 1996*b*. *The World's Youth.* Washington, D.C.

——. 1997. *Family Planning Saves Lives.* 3rd edition. Washington, D.C.

Profamilia. 1990. Encuesta de Prevalencia. Bogotá, Demografía y Salud.

RAI, U. 1992*a*. Will Bill To Ban Sex Test Control Medical Mercenaries? *Indian Express,* July 7.

——. 1992*b*. Female Infanticide Thrives in South. *Indian Express*. August 8.

Ramalingaswami, V.; Jonsson, U.; Rodhe J. 1996. The Asian Enigma. *The Progress of Nations 1996*. New York, UNICEF, pp. 11–17.

Ramanamma, A.; Bambawali, U. 1980. The Mania for Sons: An Analysis of Social Values in South Asia. *Social Science and Medicine*, Vol. 14B, No. 3, pp. 107–10. Quoted by Lakshmi Lingam, Sex-Detection Tests and Female Foeticide: Discrimination Before Birth. *The Indian Journal of Social Work*, Vol. L11, 1991, No. 1, pp. 12–19.

Ramos S.; Viladrich. 1994. *Abortos Hospitalizados: Entrada y Salida de Emergencia, en Atención Hospitalaria y Costos del Aborto*. Bogotá, Universidad Externado de Colombia.

Rawls, J. 1971. *A Theory of Justice*. Cambridge, Massachusetts, Belknap Press of Harvard University Press.

Riquer, F.; Saucedo, I.; Bedolla, P. 1996. Agresión y Violencia contra el Género Femenino: Un Asunto de Salud Pública. In: A. Langer and K. Tolbert (eds.), *Mujer: Sexualidad y Salud Reproductiva en México*., México, Population Council and EDAMEX.

Roberts, D. E. 1996. Reconstructing the Patient: Starting with Women of Color. In: S. M. Wolf (ed.), *Feminism and Bioethics*. New York, Oxford University Press, p. 116.

Robey, B.; Rutstein, S.; Morris, L.; Blackburn, R. 1992. The Reproductive Revolution: New Survey Findings. *Population Reports*, Series M, no. 11.

Roe, Bong Mi. 1987. The 'Sun' of the Family. *The Hindustan Times*, February 22.

Ruiz Gonzalez, E. (ed.). 1995. Temas de Pediatría. Edición Especial. Sistema Nacional de Síndrome Del Niño Maltratado. Defensoría del Pueblo – Colombia. IX Congreso Colombiano de Pediatria, Barranquilla, 1995.

Russell, C. 1987. Ageing as a Feminist Issue. *Women's Studies International Forum*, Vol. 10, No. 2, pp. 125–32.

Saavy, 1985. Bombay, April.

SACHS, A. 1994. Men, Sex, and Parenthood in an Overpopulated World. *World Watch*, Vol. 7, No. 2, pp. 12–19.

SANTOW, G.; BRACHER, M. 1992. Correlates of Hysterectomy in Australia. *Social Science and Medicine*, Vol. 34, No. 8, pp. 929–42.

SCHEI, B. 1991. Sexual Factors and Pelvic Pain: a Study of Women Living in Physically Abusive Relationships and of Randomly Selected Controls. *Jour. Psychosom. Obstet. Gynaecol.*, Vol. 12 (Suppl.), pp. 99–108.

SCIARRA, J. 1994. Infertility: an International Health Problem. *International Journal of Gynecology and Obstetrics,* Vol. 46, pp. 155–63.

SENANAYAKE, P.; LADJALI, M. 1994. Adolescent Health: Changing Needs. *International Journal of Gynecology and Obstetrics,* Vol. 46, pp. 137–43.

SHRADER-COX, E. 1992. Developing Strategies: Efforts to End Violence against Women in Mexico. In: M. Schuler (ed.), *Freedom from Violence: Women's Strategies from Around the World.* Washington, D.C., OEF International.

SILVERMAN, W. A. 1991. Neutral versus Aggressive Premature Baby Care. *New York Times*, Friday, October 18.

SIRISAMBHAND, N.; GORDON, A. 1987. Rural Women and Changes in Work Patterns: The Impact of a Reservoir and Transportation Networks on Three Thai Villages. In: N. Heyzer (ed.), *Women Farmers and Rural Change in Asia: Towards Equal Access and Participation.* Kuala Lumpur, Asian and Pacific Development Centre, pp. 313–52.

STEVENS, E. 1973. Marianismo: The Other Face of Machismo in Latin America. In: A. Pescatello (ed.), *Female and Male in Latin America.* Pittsburgh, University of Pittsburgh Press.

SWISS, S.; GILLER, J. E. 1993. Rape as a Crime of War. *Journal of the American Medical Association*, Vol. 270, pp. 612–5.

SYMONDS, P. 1995. *Journey to the Land of Light: Birth from the Perspective of Hmong in Maternity and Reproductive Health in Asian Societies.* Sydney, Australia, Harwood Academic Publishers.

TEREFE, A.; LARSON, C. P. 1993. Modern Contraception Use in Ethiopia: Does Involving Husbands Make a Difference? *Am. Jour. of Public Health*, Vol. 83, No. 11, pp. 1567–71.

TOUBIA N. (ed.). 1994. *Arab Women: A Profile of Diversity and Change*. Cairo, Population Council.

——. 1995. *Female Genital Mutiliation: A Call for Global Action*. New York, RAINBO. (In English, Arabic and French.)

TURMEN, T. 1995. *A Framework for Action*. Meeting on Reproductive Health in Emergency Situations 5–6 April 1995. Geneva, World Health Organization.

UNITED NATIONS (UN). 1948. Declaration of the Rights of the Child. *Universal Declaration of Human Rights*. New York, United Nations, A/810.

——. 1991. *The World's Women 1970–1990, Trends and Statistics*. New York, United Nations.

——. 1993*a*. *Report on the Abuse of Human Rights in the Territory of the Former Yugoslavia*. Geneva, United Nations. (Doc. E/CN 4/1993/50.)

——. 1993*b*. The World Conference on Human Rights. *The Vienna Declaration and Programme of Action*. June.

——. 1994*a*. *Final Report of the Commission of Experts Established Pursuant to Security Council Resolution (780)*. Copenhagen, United Nations Information Centre for the Nordic Countries.

——. 1994*b*. *Report of the International Conference on Population and Development*. New York. (A/Conf.171/13.)

——. 1994*c*. *World Abortion Policies*. New York, United Nations.

——. 1995*a*. *Fourth Conference on Women, Draft Platform of Action, Beijing*.

——. 1995*b*. *Report of the Fourth World Conference on Women*. New York, United Nations. (A/CONF.177/20.)

——. 1995*c*. *Population and Development*. Vol. 1, Programme of Action adopted at the International Conference on Population and Development, Cairo, 5–13 September 1994., New York, Department for Economic and Social Information and Policy Analysis, United Nations.

——. 1995*d*. *Report of the Committee on the Rights of the Child* (8th Session). (Doc. CRC/C/15/Add.34, para. 14.)

——. 1995*e*. *The World's Women, 1995: Trends and Statistics*. Social Statistics and Indicators Series K, No. 12, New York, United Nations.

——. 1996*a*. *Population and Women. Proceedings of the United Nations Expert Group Meeting, Gabarone, Botswana 22–26 June 1992*. ST/ESA/SER.R/130. New York, United Nations.

——. 1996*b*. *International Human Rights Instruments*. HRI/Gen/1/ Rev.2.29 March 1996, CEDAW General Recommendation No. 14. Female Circumcision, 108–9.

——. 1996*c*. *The Beijing Declaration and the Platform for Action, Fourth World Conference on Women, Beijing, China, 4–15 September 1995*. New York, United Nations Dept. of Public Information.

——. 1997*a*. *Report of the Committee on the Rights of the Child* (15th Session). (Doc. CRC/C/15/Add.75, para. 45.)

——. 1997*b*, *Report of the Committee on the Elimination of Discrimination Against Women* (16th Session). (Doc. A/52/38 (Part I), par. 147 (1997).)

——. 1997*c*, *Report of the Committee on the Elimination of Discrimination Against Women* (17th Session). (Doc. A/52/38/Rev. 1 (Part II), par. 108 (1997).)

UNAIDS (Joint United Nations Program on HIV/AIDS). 1996. *Fact Sheet*. December, Geneva, United Nations.

UNESCO. 2000. *World Education Report 2000*. Paris, UNESCO Publishing.

UNDP (United Nations Development Program). 1994. *Human Development Report 1994*. Oxford, Oxford University Press.

——. 1995. *Human Development Report 1995*. Oxford, Oxford University Press.

UNHCR (United Nations High Commissioner on Refugees). 1995. *Information Notes on Former Yugoslavia*. No. 1/1995. Zagreb, UNHCR Office of the Special Envoy for Former Yugoslavia, External Relation Unit.

UNI Report. 1994. Sex Test Kills 3,500 Foetuses in Jaipur. *Indian Express*, New Delhi, 31 January 1994.

UNICEF. 1996. *Progress of Nations*. New York, UNICEF.

——. 1997. *The State of the World's Children*. New York, Oxford University Press for UNICEF.

UNITED NATIONS POPULATION FUND. 1997. *The State of World Population*. New York, UNFPA.

UNITED STATES NATIONAL INSTITUTES OF HEALTH. 1992. *Opportunity for Research on Women's Health*. NIH Publication No. 92-3457. Washington, D.C., United States Department of Health and Human Services.

VALDES, T.; GOMARIZ, E. 1993. *Mujeres Latinoamericanas en Cifras. Argentinas*. Spain, FLACSO/Instituto de la Mujer.

VAN DEN HOONAARD, D. K. 1997. *Older Women's Experiences of Widowhood*. Frederickton, Third Age Centre, St Thomas University.

Velasquez Rodriquez Case (Honduras). 1988. Inter-American Court of Human Rights (ser.C) at 92, at par. 166.

VISARIA, L.; VISARIA, P. 1995. India's Population in Transition. *Population Bulletin*, Vol. 50, No. 3, Washington, D.C., Population Reference Bureau, October.

WARE, H. 1981. *Women, Demography and Development*. Development Studies Centre Teaching Notes 3. Canberra, Australian National University.

WORLD HEALTH ORGANIZATION (WHO). 1946. *Preamble to the Constitution of the World Health Organization*. Official Records of the World Health Organization 100 (July 1946).

——. 1992. *Recent Advances in Medically Assisted Conception*. WHO Technical Report Series 820, Geneva, World Health Organization.

——. 1994*a*. *Abortion: A Tabulation of Available Data on the Frequency and Mortality of Unsafe Abortion*. 2nd Edition, WHO/FHE/MSM/93.13, Geneva.

——. 1994*b*. *Investing in Women's Health: Central and Eastern Europe*. Copenhagen, WHO Regional Office for Europe.

——. 1995*a*. *Biennial Report 1994–95*. Geneva, World Health Organization.

——. 1995*b*. *Facing the Challenges of HIV/AIDS/STDs: A Gender-based Response*. Kit & SafAIDS, Geneva, World Health Organization.

——. 1995*c*. *Women's Health*. WHO position paper for Fourth World Conference on Women (Beijing, 1995), Geneva, World Health Organization.

——. 1995*d*. *The World Health Report 1995*. Geneva, World Health Organization.

——. 1996. *Female Genital Mutilation*. Report of a WHO Technical Working Group, Geneva, 17–19 July, 1995, WHO/FRH/WHD/96.10.

——. 2000. *WHO Information Fact Sheets*. No. 241. June. http://www.who.int/inf-fs/en/fact241.html

WHO; UNICEF. 1996. *Revised 1990 Estimates of Maternal Mortality: A New Approach by WHO & UNICEF*. Geneva, World Health Organization, April.

WOLF, S. M. (ed.). 1996. *Feminism and Bioethics: Beyond Reproduction*. New York, Oxford University Press.

——. 1994. Women's Voices '94 Declaration, made before the 1994 International Conference on Population and Development. In: L. A. Mazur (ed.), *Beyond the Numbers: A Reader on Population, Consumption, and the Environment*. Washington, D.C., Island Press.

WORLD BANK. 1993. *World Development Report 1993: Investing in Health*. Oxford, World Bank/Oxford University Press.

WORLD FEDERATION OF MENTAL HEALTH. 1993. *Proceedings of Annual Conference*. WFMH, Japan.

WORLD RESOURCES INSTITUTE. 1994. *World Resources 1992–1993*, Washington, D.C., World Resources Institute.

X and Y vs. *The Netherlands*. 1985. European Court of Human Rights (ser. A).

ZULFIKAR, M. 1995. *Women in Development: A Legal Study*. Cairo, UNICEF.